FOOTPRINTS

Two Hundred Billion and Counting

TED FOLKERT

outskirts press

Footprints
Two Hundred Billion and Counting
All Rights Reserved.
Copyright © 2023 Ted Folkert
v1.0 r1.0

The opinions expressed in this manuscript are solely the opinions of the author and do not represent the opinions or thoughts of the publisher. The author has represented and warranted full ownership and/or legal right to publish all the materials in this book.

This book may not be reproduced, transmitted, or stored in whole or in part by any means, including graphic, electronic, or mechanical without the express written consent of the publisher except in the case of brief quotations embodied in critical articles and reviews.

Outskirts Press, Inc.
http://www.outskirtspress.com

ISBN: 978-1-9772-5916-5

Cover Photo © 2023 www.gettyimages.com. All rights reserved - used with permission.

Outskirts Press and the "OP" logo are trademarks belonging to Outskirts Press, Inc.

PRINTED IN THE UNITED STATES OF AMERICA

Table of Contents

Preface i
Introduction vii
Chapter 1: Voices of the Past 1
Chapter 2: Other Well-known Voices 14
Chapter 3: Recent Common Good Voices 19
Chapter 4: More Voices for the Common Good 28
Chapter 5: Entertainers Opinions 41
Chapter 6: Income & Taxation 55
Chapter 7: Leveling the Playing Field 64
Chapter 8: Political Campaigns 69
Chapter 9: Measures for Political Equality 77
Chapter 10: Footprints on a Struggling Planet 81
Chapter 11: Pros and Cons of Equality 90
Chapter 12: What's it all About? 99
Chapter 13: The Impact of the Cost of War 103
Chapter 14: The Path to Power and Privilege 107
Chapter 15: Political Opinions 113
Chapter 16: Measures for Enhancing Equality 117
Chapter 17: Democratic Governments 122
Chapter 18: Greed Revisited 134
Chapter 19: Governmental Choices 140
Chapter 20: Alternative Campaign Finance 145
Chapter 21: Taxation – the only path to equality 148
Chapter 22: Funding of Political Parties 153

Chapter 23: Income Tax Data 157
Chapter 24: The Ultimate Challenge 162
Concluding Comments 167
Addendum 173
Notes and References 179

Preface

WE ALL LEAVE our footprints in the sand, so to speak. We all have an impact on the health of this lovely rock we inhabit - this amazing Planet Earth. Our scientists tell us that this amazing rock encircles the sun each year at a speed of about 67,000 miles per hour and maintains an average temperature of about 60 degrees Fahrenheit, a mind-boggling thought, no less. It maintains climatic conditions which we humans require and has done so for millions of years. It is the only planetary body of which we are aware that supports life as we know it. The other planets in our universe: Mercury, Venus, Mars, Jupiter, Saturn, Uranus, and Neptune are not known to contain life-sustaining conditions as we know and require them for human habitation.

Scientific researchers have estimated that homo-sapiens have existed as early as 50,000 B.C. E., some say as early as 200,000 B.C.E. They believe that 7.8 billion of us now exist on Planet Earth and that perhaps 117 billion have existed hereon over these 200,000 years. Assuming that most everyone has two feet it represents more than 200 billion unique footprints on the planet.

Our scientists tell us that we humans are not treating this planet with the respect necessary to assure the support of human life indefinitely. They tell us that we aren't protecting the air we breathe, that which we depend upon for survival. They tell us that we aren't protecting the oceans and the fresh water sources that we depend upon for survival. They tell us that we are not protecting the croplands that we depend upon for survival. They tell us that our actions are causing a heating of the planet due to our fossil fuel emissions which have created a layer of carbon dioxide around the planet that is increasing year by year. They tell us that this will cause the oceans to rise and shrink the habitable land area of the planet, a devastating situation for some areas, initially some island countries and eventually every landmass on the planet.

Scientists tell us that human domination of the Earth's ecosystems has caused an extinction of ten percent or more of animal species thus far and that we risk the extinction of thirty percent of animal species over the next hundred years. They tell us that human activity has had a significant negative impact on the surface of the planet and that the increase of carbon dioxide in the atmosphere has increased by fifty percent since the Industrial Revolution of the eighteenth and nineteenth centuries.

Many of our actions which are said to be causing this potential demise are considered to be essential for our survival. These are our footprints. We may be unable to change the impact on those provisions which we consider essential. We certainly can, however, change some of the nonessential practices for personal comfort, pleasure, fame, and fortune which drive the quest for power and privilege. Therein must lay an enormous and unnecessary impact on the environment which we can change. But before we can change this thoughtless potential devastation of planetary living conditions, we must identify the causes and agree to take the steps necessary for survival. It must be done worldwide. We must do it for the common good of mankind, present and future.

Common good is a term popular for writers, pleasant to the ears, and one which most of us agree upon and attempt to pursue. Who could disagree with the common good as the reining principle of democracy, the basic measure of equal rights and equality of opportunity? It isn't a worn out cliché. No, it is a condition, perhaps an oath, which should prevail for all people and for all time. But do those of us who focus entirely on wealth, personal pleasure, and image enhancement consider the common good as our guiding principle?

That doesn't seem possible. It doesn't seem that the two principles of wealth enhancement and the common good "go together like a horse and carriage" as the song line states about "love and marriage."

Our political leaders and office-seekers speak of the common good as an achievement in and of itself. They emphasize equal rights and economic equality of opportunity for all as their primary interest. That intent sounds encouraging as a principle for humanity, however, what many of them seem to focus on and seek for themselves are those advantages of hierarchy which are not common, not universally shared, not easily obtained positions – power and privilege.

The quest for power and privilege drives ambition, dedication, perseverance, persistence, determination, and tenacity. It emphasizes the importance of education, which becomes imperative in the advancement of income and wealth. Income and wealth enable prosperity and fame. The desire for prosperity and fame drive discovery, invention, development, and scientific advancement.

The yearning, seeking, and achieving of power and privilege is justified as necessary for the ability to distribute equality of opportunity to all, the proclaimed right for all mankind, the essential condition to enable positive economic performance in providing the expanding demand for food, shelter, clothing, healthcare, defense, and education for our ever-growing world population.

Of course, food and shelter are the most basic needs of humanity, but all the other elements of prosperity are considered imperative in providing this ever-growing demand.

We call this process capitalism, a form of government to which we are willfully consigned as the most logical path to equality of opportunity. But a recent lingering question which weighs upon this economic system implies a threat for mankind: "Can humanity survive under capitalism"? We assume this to be inevitable as we go about our lives. However, some of our leading scholars don't believe that future generations can retain the lifestyles we enjoy and enable humanity to survive. Should we consider the plight of our children or our children's children? Should we merely do what serves our needs and pleasures regardless of the consequences for those who follow?

These rhetorical, yet pertinent, questions set the stage for a conversation about economic and socialistic matters, a topic which should head the list of scholastic endeavors throughout the future educational experience. Perhaps these should top the list of the most important issues of the day rather than the incessant struggle to achieve status with inclusion on the list of millionaires or billionaires.

Some of our educators and prognosticators say that we have the "best government that money can buy" – a cliché which is difficult to counter considering our campaign finance laws that are proposed, promoted, and enacted by legislators influenced by the power of money in campaign finance.

The implication is that government is not elected, it is bought. I suppose it can be considered to be bought if we acknowledge the lack of limitations on campaign finance. Although some politicians and elected officials deny that government is bought by campaign contributions, they seldom, if ever, resist them when their own election campaign comes about. Campaign finance not only loses impotence as an issue then, it becomes imperative since those seeking office are unlikely to get elected or reelected without the support of their campaign finance sources, and they cannot get that support without an adequate allegiance to democratic principles while simultaneously making some commitments of cooperation, even if it they believe it to be a challenge to their proclaimed ideals and principles of governance.

The troubling economic evidence today suggests that the gap between the rich and everyone else has reached a dangerous proportion for the economic health of the nation and probably the world. This is not casual commentary by the working class or the desperately poor. This is the common belief among our leading economists and sociologists who monitor the levels of income and wealth equality and advise our leaders regarding economic matters that endanger the health of our economy. Some of our billionaires even suggest that they should be taxed at a higher rate than at present. This is of course a common belief amongst the proletariat.

This dilemma sets the stage for a continuing conversation about the practicality or legitimacy of being ruled by elected officials who are indebted to their campaign financiers. How does this process square with the phrase which we all cherish, "the common good?" That would seem to infer its availability to all, including most of us who are considered commoners. This begs the question: "Are economic matters legislated for the benefit of all or legislated for the benefit of those who finance political campaigns?" The answer to this dilemma can engender an ongoing debate among the rich, the poor, and the scholars of economic matters. It seems imperative to steer the benefit toward the populous if we wish to retain our system of quasi-democracy, although the prospects under our system of capitalism seem to make that either unlikely or impossible.

This writing supports the theory that this drive for power, privilege and pleasure, although considered by those who prosper as essential for a healthy system of free enterprise, provides a serious threat to sustainability of the elements supporting human life on the planet. It is hopeful that the consensus of the scientific community can serve to convince the populous that this threat, this trail of evidence, these footprints in the sand, can instill action to lessen the impact of this imminent threat to humanity. And it is hopeful that we will come to realize that the time for action to lessen this impact upon future generations is now.

Introduction

POWER AND PRIVILEGE have never been and are unlikely to ever be shared equally by everyone. If so, there wouldn't be any such thing as power and privilege. That would seem unrealistic, perhaps averse to principles of free enterprise, the so-called basis of capitalism. Perhaps neither our founders nor the political leaders who have followed intended a high level of equality of opportunity. It seems that they assumed power and privilege and pursued them and practiced them. Perhaps they assumed that too much equality could destroy the motivations for progress in our economic system, which would be difficult to control without some method of pursuit and attainment of power and privilege.

Protest of the perceived lack of equality of opportunity provided by our system of free enterprise is typically expounded and proclaimed by the proletariat, those who lack power and privilege. Of course, power and privilege are alleged to be rewards for hard work, diligence, or perhaps birthright by those in positions of control, those financially successful through inheritance or their own achievements, those who are enriched by the power of money. They become and remain powerful as their wealth enables them inordinate control of the political system.

Free enterprise, which supposedly enables and determines power and privilege, is not actually free. Effective use thereof is determined by money, the medium of exchange which accumulates in excess to those who have attained it through hard work or been awarded it through birthright, or good fortune. This may seem like a vicious circle – such as, you can have it if you already have it – but that seems to be the way our system determines and administers power and privilege in general.

Making the case for equality of opportunity is not a novel idea or a new concept, but one which escapes reality due to the power of money in political campaign finance – the immovable blockade to representative governance – that inoperable cliché defined as government of, by, and for the people, as stated in the Preamble to The Constitution of the United States of America. It should be noted that such a reference does not read "… government of, by, and for all the people," and it doesn't identify which people. No, it simply states "government of, by, and for the people," the interpretation of intent being at the discretion of the user as it best serves the cause in question.

Power and privilege are defining factors in the distribution of wealth and the necessities of life in a system of free enterprise, a system to which we prescribe as a necessity for enabling equality of opportunity.

Protest is the defining factor of expressing the inequality in such distribution, an inherent method for those without power and privilege to argue for or attempt to level the playing field, so to speak, a method of achieving a measure of democracy without which equality of opportunity could not serve society in a fair and equitable manner.

What is it that drives us to pursue success in terms of ownership of money, property, or rights to the use or access of either? Is success all about money? Is it about the things money will buy? Is it about power and influence? Is it about comfort and pleasure? Or is it merely about survival? Is it keeping score in a game of chance? Is it a game of winner-take-all? Is the one with the most money or most toys when the game ends the winner? Is it a game that never ends?

Of course there are no direct answers for these rhetorical questions. It depends upon one's vantage point, where one is on the income and survival scale and what goals of satisfaction one cherishes.

Actually, success can mean any or all of the above depending upon one's place on the ladder between poverty and opulence, between complacency, success, or fruitless struggle.

Capitalism, the economic term which we use to describe our medium of exchange, seems to be the economic system followed in some form by most civilizations which value democratic forms of government, those which profess to be established and operated of, by, and for the governed, rather than other forms of ownership and exchange such as monarchy, with one leader, plutocracy, with more than one, or socialism or communism, whereby governing and ownership are thought to be shared by the populace – although, such method of governance never seems to survive without strong leadership.

The primary argument for supporting capitalism is that it provides the basis for universal progress through individual ambition and desire for compensation for one's individual efforts rather than common distribution of goods and services regardless of personal contribution toward the general welfare.

It seems that a lack of compensation based upon one's individual efforts would be a deterrent to any combined effort in providing the necessities of life for all of world's inhabitants. Who would do their fair share or more than the next person? Who would assign the duties required by each individual? Who would invent, establish, organize, or administer? Who would monitor the performance of every individual?

All of such questions are rhetorical. Such a system would be difficult-to-impossible to facilitate, administer, or monitor. It seems apparent that the best way an economy can perform in any means of equality would be some form of capitalism, although such a system in practicality fails to provide meaningful equality of opportunity. In fact it is averse to such for us due to our system of taxation which provides the funds necessary to administer our government. Taxation is slanted favorably toward those with wealth – not toward those who lack power and privilege, the working class, the proletariat.

So, it seems obvious that the lack of any more democratic method of the distribution of goods and services which could function within a reasonable means of equality of opportunity would be difficult or impossible to create or administer. If we accept such as reasonable support for a system of capitalism, then we can proceed with consideration of the methods of improving the system to better serve all the people rather than those who prosper from advantageous positions of good fortune maintained through the power of money, either earned through enhanced or unfair compensation or from inherited wealth previously attained through enhanced or unfair compensation.

What drives the value of real property, for example? Real property values generally increase through demand, restrictions, scarcity, replacement-cost, or capital improvements of the given property or the particular area in general. It often increases through restrictions based upon wealth, national origin, or race. For example, property values have not increased at the same pace in Inglewood, California as they have in Beverly Hills, California. Is that because the land is more productive or of a more favorable strategic locale? No, it has more likely increased because of restrictions based upon race and national origin, or because those with wealth choose to live separately from those without.

Property values probably have not increased as fast or as much in rural New York as they have in New York City. They probably have not increased as much in Downtown Kansas City as they have in the Country Club Plaza or Overland Park in the Kansas City area. Is that because the land is more productive or of a more favorable strategic locale? No, it has more likely increased because of restrictions based upon race or national origin, or because those with wealth want to live separately from those without and those with wealth want to live close to the center of American capitalism where they can continue to prosper by dealing directly with their counterparts in the economy, a system which either intentionally or unintentionally excludes or hampers the ability of those of us who lack adequate wealth or political power to participate.

Political leaders who profess "common good" principles have

suggested, promoted, and enacted legislation of various types and proportions to sort of level the playing field over the years, however, such measures have usually been ignored, gutted of enforcement provisions, revised with pertinent elements extracted, limited by special interests, or rendered impractical by legal challenges. And, who is better able to afford legal challenges than those with excess wealth garnered through the system which they wish to maintain.

Equality of opportunity, the phrase we all cherish, is difficult or impossible to achieve in practice. This phrase, along with the phrase "where there's a will there's a way," is more effectively utilized by the rich and powerful than by the powerless proletariat who lack the time or wealth to attempt to change or challenge the distortions of the proclaimed intent. Perhaps the phrase suggested by some should be: "where there's wealth there's a way."

Our libraries and book shelves offer many volumes concerning this subject matter. We don't all follow, understand, or admire the same opinions, pros, and cons of our economic system. But is it a system? Perhaps it is more like rules of a game which are made up as the game progresses. The ability to utilize whatever system shall prevail is not a level playing field by any means of imagination. One's ability to participate is determined by the very thing which any description would classify as power and wealth.

It is a well-accepted concept among the scholastic and financial world that the desire for power and privilege drive economic growth and that the absence of such would be a limiting factor in sustaining an adequate and consistent supply of goods and services to meet demand, the primary function of and necessity for a stable economy.

Of course, in discussing this imperative subject matter we can't ignore the greatest obstacle to equality of opportunity – the existing and widening distribution of income between the wealthy and privileged class and the wage-earning workers.

In understanding power, privilege, and protest, we are blessed with the contributions of many brilliant authors, poets, song writers, musicians, and story-tellers from the past and the present. The tributes could be endless and cover centuries of literature. Many of us have our favorites - those who we feel excel at exemplifying the human passions, hopes, failures, and disappointments which life bestows upon us.

This writing intends to offer contributions and opinions of some of our popular writers and performers of literature, poetry, and song who have dedicated their time and efforts in defining, describing, and supporting the societal effects of power and wealth. These contributions deserve to be acknowledged as important measures in the ongoing challenge of attaining common good as a defining principle in governmental laws, rules, regulations, and practices.

The need for these voices intensifies as we the workers slide quietly into the abyss of a struggling economy, with the playing field tilted evermore to the rich and powerful. The ability to maintain a satisfactory standard of living will become increasingly more difficult for wage earners as the slide continues due to the lack of funding and leadership to express a unified resistance and demand for change.

And the debated question remains: will a system of free-enterprise survive and provide prosperity for all if the workers have no money in their pockets? If not, what will drive the economy, who will buy the goods and services, how will the business owners and the financiers survive without a healthy marketplace?

This seems to be a conversation of opinions slanted to one's place on the hierarchy of income and wealth, a conversation which has no consensus, and a conversation which should be prioritized to enhance equality of opportunity, which is essential for democratic principles to prevail. Unfortunately recent actions by our wealthy leadership have followed the lead of those with the power of money, a power which isn't and never could be defined as equal rights or equal opportunities.

Thus far, the irresolvable question remains: How can those without power and wealth have their voices heard – how can the playing field of this game of chance be leveled to serve the entire populous in an equitable manner instead of being tilted favorably for those with power and privilege – that condition we erroneously call equal rights? In our system, rights are not equal. They are determined by political influence, which is determined by campaign finance, which is provided primarily by those who own the wealth and who are motivated to retain control of wealth distribution through favorable definitions of taxable income and the taxation thereof.

Footprints - two hundred billion and counting - have presented the challenge. We the people, now with about fifteen billion footprints, must provide the solution. The outcome depends upon making the pertinent decisions and instituting the pertinent actions, an immense task which becomes more imminent and more challenging each year.

CHAPTER 1

Voices of the Past

MANY WELL-KNOWN SCHOLARS, poets, and musicians have leant their voices to the challenge of promoting common good and supporting the economic policies of free-enterprise and equality of opportunity. Some of such works describe the struggles for the achievement of power and wealth. Some question the social practicality of such a struggle. Others suggest paths to progress in achieving and maintaining a distribution of power and wealth which provide equality of opportunity through fair and equitable income and taxation.

The voices addressing power and privilege are too numerous to mention in a limited framework, but many stand out as having made exceptional contributions to the common cause. They include such earlier contributors of personal note as Homer in the 8th century BCE, from thousands of years ago, and Leo Tolstoy and Victor Hugo in the 19th century, from some 200 years ago.

More recently, those of literary fame include: Ruth Bader Ginsburg, Woody Guthrie, James Joyce, Leonard Cohen, Bob Marley, John Lennon, Arlo Guthrie, Bert Bacharach, Robert Reich, Noam Chomsky, Al Gore, Al Franken, and many others.

Writers of a more scholastic nature include: Harriet Beecher Stowe, Anna Elizabeth Dickenson, Thomas Friedman, Paul Krugman, Thomas Sowell, Peter F. Drucker, William Greider, Charles H. Ferguson, Upton Sinclair, Joseph E. Stiglitz and many others.

And, of course, William Shakespeare, the inimitable playwright, enters into any discussion of the common good with his poetic voice.

Many notables of the music world have been dedicated contributors to the voice of the common good. They include: Bob Dylan, Paul Simon, Neil Young, Arlo Guthrie, Billy Joel, Willie Nelson, The Beatles, Johnny Cash, Stevie Wonder, and Sam Cooke, among many others.

The following sections provide a brief review of some of the noteworthy contributions of our scholars, poets, and musicians as they have leant voices to the support, success, or failure of measures striving for equality of opportunity in our American experiment of quasi-democracy.

Homer - Greek Poet[1]

As a personal note, my father's name was Homer Theodore Newcomb, his birth-name in 1912. His parents were insightful in acknowledging history by choosing their son's name from that of the historically famous poet and orator, Homer, and the ground-breaking former U.S. president of the era, Theodore Roosevelt.

Homer, a Greek poet who lived during the years of about 800 BCE to 700 BCE, is attributed to be the author of the famous epic poems *The Iliad* and *The Odyssey*, masterworks of world literature which have had a lasting presence in classic literature. Very little is known about the alleged author of these classics, Homer, who was believed by some to have been blind. The poems are said to have been recited regularly at festivals in Athenian culture over many years.

Plato, of the 5th century BCE, a student of Socrates and teacher of Aristotle, tells us that in his time many believed that Homer was the educator of all Greece. "Since then, Homer's influence has spread far beyond ...." wrote Werner Jaeger in "Paideia: The Ideals of Greek Culture." He was right. "*The Iliad* and *The Odyssey* have provided not only seeds, but fertilizer, for almost all the other arts and sciences in Western culture. For the Greeks, Homer was a godfather of their national culture, chronicling its mythology and collective memory in rich rhythmic tales that have permeated the collective imagination." [1]

Homer's real life may remain unverifiable, as there are varying opinions regarding the exact time of his life and his oral or written contributions to our culture, but the very real impact of the works attributed to him continue to illuminate our scholarly world today.

The Iliad[2]

The Iliad, an ancient Greek epic poem of more than hundred thousand words and hundreds of pages, dated to have been from about the 8th Century BC, has been attributed to Homer. It tells of the battles and events during the weeks of a quarrel between King Agamemnon and the warrior Achilles. Although the story covers only a few weeks in the final year of the war, it includes many of the Greek legends about the siege and the cause of the war. As these events are described in the poem it tells the tale of the Trojan War.

The main theme of the poem is that of war and peace, and the poem is essentially a description of war and fighting. There is a sense of horror and futility built into Homer's chronicle, and yet, posed against the viciousness, there is a sense of heroism and glory that adds a glamour to the fighting: Homer appears both to abhor war and to glorify it. Frequent similes tell of the peacetime efforts back home in Greece, and serve as contrasts to the war, reminding us of the human values that are destroyed by fighting, as well as what is worth fighting for.

The concept of heroism, and the honor that results from it, is also one of the major currents running through the poem. Achilles in particular represents the heroic code and his struggle revolves around his belief in an honor system, as opposed to Agamemnon's reliance on royal privilege. But, as fighter after heroic fighter enters the fray in search of honor and is slain before our eyes, the question always remains as to whether their struggle, heroic or not, is really worth the sacrifice.

Perhaps this begs the question about all wars. What causes are worth fighting for and are they really worth the sacrifice? These are obviously rhetorical questions. The story exemplifies power and privilege and the struggle for a resolution of right and wrong through a perceived system of honor and sacrifice. Any didactic meaning is left to the discretion of the reader.

Assuming that beating a child into obedience is often an unsuccessful, or at least an unpleasant or undesirable corrective measure, how can warfare beat other countries into obedience in becoming better world citizens with democratic principles, fairness and compassion for other countries, and universal equality of opportunity, especially in consideration of the domineering impact of authoritarian rule in many countries of the world?

And there is the belief of many historians that many wars were unnecessary for any other reason than power struggles between nations and leaders of nations. And the rhetorical but vital question remains – how many of the wars between countries or leaders would not have occurred if the leader who declared war was compelled to lead the troops into battle while sending other people's children to fight the battles.

The Odyssey[3]

The Odyssey, also an epic poem of more than hundred thousand words and hundreds of pages, is also traditionally attributed to Homer. The poem is the story of Odysseus, king of Ithaca, trying to get home after the Trojan War. After the war itself, which lasted ten years, his journey lasts for ten additional years, during which time he encounters many perils and all his crewmates are killed. In his absence, Odysseus is assumed dead, and his wife Penelope and son Telemachus must contend with a group of unruly suitors who compete for Penelope's hand in marriage. On his return, Odysseus is recognized only by his faithful dog and a nurse. With the help of his son, Telemachus, Odysseus destroys the insistent suitors of his faithful wife, Penelope, and several of her maids who had fraternized with the suitors, and reestablishes himself in his kingdom.

Ezra Pound stressed Homer's perennial freshness in his comment in the 2013 Edition of the Iliad and the Odyssey – "that after more than 2,500 years the Iliad and the Odyssey still speak to us, unforgettably, about nearly everything in life that matters." This publication states that: "It can be clearly seen that the audiences then did not differ much from our own after some three thousand years."

James Joyce, Irish Novelist

James Augustine Aloysius – 1882 to 1941 - was an Irish novelist, short story writer, poet, teacher, and literary critic. He contributed to the modernist avant-garde movement and is regarded as one of the most influential and important writers of the 20th century.

Ulysses[4]

James Joyce's novel, *Ulysses*, of the 20th Century, tracks some of the features of *The Odyssey*, although it focuses on events of a single day rather than a ten year span of time. Since its publication, the book has attracted controversy and scrutiny, but has been regarded by many as one of the greatest literary works in history.

While Joyce's novel takes place during one ordinary day in early 20th-century Dublin, in Homer's epic, Odysseus, "a Greek hero of the Trojan War ... took ten years to find his way from Troy to his home on the island of Ithaca". Furthermore, Homer's poem includes violent storms and a shipwreck, giants and monsters, gods and goddesses, a totally different world from Joyce's. Joyce's character, Leopold Bloom, "a Jewish advertisement canvasser," corresponds to Odysseus in Homer's epic; Stephen Dedalus, the hero also of Joyce's earlier, largely autobiographical *A Portrait of the Artist as a Young Man*, corresponds to Odysseus's son Telemachus; and Bloom's wife Molly corresponds to Penelope, Odysseus's wife, who waited 20 years for him to return.

T. S. Eliot, poet, essayist, publisher, playwright, literary critic and editor – 1888 to 1965, said of *Ulysses*: "I hold this book to be the most important expression which the present age has found; it is a book to which we are all indebted, and from which none of us can escape." He went on to assert that Joyce was not at fault if people after him did not understand it: "The next generation is responsible for its own soul; a man of genius is responsible to his peers, not to a studio full of uneducated and undisciplined coxcombs." [4]

The book was developed into a television series of 31 episodes in the 1980's and became a popular attraction for the television audience as it enhanced awareness of its significance in American culture and exemplified the tragedies of warfare and its ultimate impact on humanity.

Leo Tolstoy – Russian Playwright and Novelist[5]

Lev Nikolayevich Tolstoy, Leo Tolstoy, lived from 1828 to 1910. He was a Russian playwright and novelist regarded by many as one of the greatest authors of all time. He believed that the aristocracy was a burden on the poor, and that the only way to live together is anarchism. He opposed private land ownership and the institution of marriage, and valued chastity, ideals also held by the young Mahatma Gandhi of India, born 1869 and assassinated in 1948, during much of the same time period of Tolstoy's life. Tolstoy's passion from the depth of his austere moral views is reflected in his classic novel, *War and Peace*.

War and Peace[6]

War and Peace is an epic novel about Russian society between 1805 and 1815, just before and after the Napoleonic invasion. Considered one of the greatest books ever written, it contains 559 characters, commemorates important military battles and portrays famous historical personalities, but it's main theme is the chronicle of the lives of two main aristocratic Russian families, the Rostov's and the Belkovsky's.

The novel, considered a masterpiece by many, contains more

than 1300 pages with more than 500,000 words. During this time Alexander I, the peacemaker of Europe and champion of liberal reforms in his country, had taken liberal initiatives and was considered to have saved Russia. However, he was harshly criticized and highly praised at the same time, a situation which Oscar Wilde has described as: " ... virtually everyone – even people in advantageous or privileged circumstances – finds the living of life a worrying and difficult business most of the time."

The novel discusses the causes of warfare. It offers an analogy of cause: "When and apple falls from a tree what makes it fall? Is it gravity pulling it down to earth? A withered stalk? The drying action of the sun? Increased weight? A breathe of wind? Or a boy under the tree who wants to eat it" Nothing is the cause of it. It is just the coming together of various conditions necessary for any living, organic, elemental event to take place."[6]

Tolstoy, through the voice of his character, Prince Audrey Belkovsky, defines warfare:

"The aim of war is murder, the weapons of war are spying, treachery, and destruction of people, looting their property and stealing from them to keep the army on the road, falsehood and deceit, which go by the name of clever tactical ploys, and the moral basis of the military class is the curtailment of freedom through discipline linked to idleness, ignorance, cruelty, debauchery, and drunkenness, and in spite of all that, it's still the highest class, universally respected."

"People come together to murder one another, men get slaughtered and crippled and the services of thanksgiving are held to celebrate the killing of the vast numbers of men, and victory is proclaimed on the basis that more men were slaughtered, the greater achievement. How can God look down from heaven and listen to it all?"[6]

Leo Tolstoy was a profound social and moral thinker and one of the greatest writers of realistic fiction during his time. The son of a nobleman landowner, Tolstoy was orphaned at the age of 9 and taught mainly by tutors from countries like Germany and France. At the young age of 16, he enrolled in Kazan University but quickly became dissatisfied with his studies and dropped out soon after. After a brief and futile attempt to improve the conditions of the serfs on his estate, he plunged into the dissipations of Moscow's high society.

In the uniquely candid powerful novel Confession, Tolstoy described his spiritual unrest and started his long journey toward moral and social certainty. He found them in two principals of the Christian gospels: love for all human beings and resistance to the forces of evil. From within autocratic Russia, Tolstoy fearlessly attacked social inequality and coercive forms of government and church authority. His didactic essays, translated into many different languages, won hearts in many countries and from all walks of life, many of whom visited him in Russia seeking advice.

At the age of 82, increasingly tormented by the disparity between his teachings, his personal wealth and by endless fights with his wife, Tolstoy walked away from his home late one night. He became ill three days later and died at a remote railway station. At his death he was praised the world over for being a wonderfully moral man. That force and his timeless and universal art continue to provide inspiration today.[6]

War and Peace, Epilogue[7]

An epilogue of War and Peace discusses the dichotomies of human comprehension – "A bee has settled on a flower and stung a child. The child is scared of bees and says that bees are there to sting people. A poet admires the bee as it imbibes inside the sepals of the flower and says the bees are there to imbibe nectar inside the flowers. A beekeeper, observing that the bee collects pollen and brings it back to the hive, says that the bee is there to collect honey. Another beekeeper, one who has studied the life of the swarm more closely, says the bee collects pollen to feed the young ones and rear a queen, and the bee is there for the propagation of its species. A botanist observes the bee flying over with pollen to fertilize the pistil on a diclinous flower and sees this as the purpose of the bee. Another one, observing the tendency for plants to migrate and the bee's contribution to this process, feels able to claim this as the purpose of the bee. But the ultimate purpose of the bee is not exhausted by the first, second, or third purpose discernible by human intellect. The higher the human intellect goes in discovering more and more purposes, the more obvious it becomes that the ultimate purpose is beyond human comprehension." [7]

Human comprehension does not extend beyond observations of the interaction between the living bee and other manifestations of life. The same applies to the purposes of historical characters and nations. [7]

Discussing human comprehension further, Tolstoy tells us that during the ferment of Paris in 1789: "Throughout this twenty-year period fields go unplowed, houses are burnt down, trade flows in different directions, millions of men grow poor, get rich or migrate, and millions of good Christian folk who claim to love thy neighbor go about murdering each other." "What does it all mean? Why did it happen? What could have induced these people to burn houses down and murder their fellow creatures? What were the causes of these events? These are simple and honest questions that leap to mind when humanity comes across memorials and traditions stemming from that bygone age of turmoil."[7]

As we look back on our own recent world history we are compelled to ask the same inexplicable questions.

One can take most any message sought from Tolstoy's references to support personal opinions about the causes of such travesties and the attempts which followed to placate perceived wrongs. Which events were meant to resolve conflicts and which were power struggles or paths to fame and fortune?

Historians agree that the external activity of states and peoples when they are at odds with each other finds an outlet in warfare, and that the political power of states and peoples waxes or wanes in direct proportion to success on the battlefield.[8]

We may find it strange when history describes the process by which some king or emperor falls out with another king or emperor, raises an army, fights and wins a battle against the enemy, killing three, five, or ten thousand men, and thus subjugates a state and a whole nation running to millions; we find it hard to understand how the defeat of an army, a mere hundredth part of a nation's strength, somehow forces the whole people into submission, yet all the facts of history justify the general belief that the success of failure of one nation's army against another nation's army is the cause, or at least a major indication, of the waxing or waning of nation's power. An army wins a battle, and the winners' rights are immediately increased at the expense of the losers'. An army suffers a defeat, and the people are immediately deprived of their rights according to the magnitude of the defeat, and if defeat is total, they fall into total submission. It has been like this from ancient times, right up to the present day. [8]

Considering the historical commentary of these noted scholars whose thoughts and ideals we have cited, it seems that the question for modern society still remains – Is the price of war on those who had no choice in the waging of war, those who were forced into battle against their will, worth it? Perhaps it is for those who declare war, but perhaps not for those who lose limbs, lives, and livelihood carrying it out.

As Tolstoy talks about war: " ..... millions of good Christian folk who claim to love their neighbor, go about murdering each other."

Perhaps somethings never change when it comes to gaining or retaining power and privilege.

William Shakespeare [9]

Such writings as we have discussed above may remind us of a message from William Shakespeare in his classic play "As You like It, Act II. Scene VII," "The Seven Ages of Man":

All the world's a stage,
And all the men and women merely players;
They have their exits and their entrances,
And one man in his time plays many parts,

His acts being seven ages. At first the infant,
Mewling and puking in the nurse's arms.
Then, the whining school-boy with his satchel
And shining morning face, creeping like snail
Unwillingly to school. And then the lover,
Sighing like furnace, with a woeful ballad
Made to his mistress' eyebrow. Then, a soldier,
Full of strange oaths, and bearded like the pard,
Jealous in honour, sudden, and quick in quarrel,
Seeking the bubble reputation
Even in the cannon's mouth. And then, the justice,
In fair round belly, with a good capon lined,
With eyes severe, and beard of formal cut,
Full of wise saws, and modern instances,
And so he plays his part. The sixth age shifts
Into the lean and slippered pantaloon,
With spectacles on nose and pouch on side,
His youthful hose, well saved, a world too wide
For his shrunk shank, and his big manly voice,
Turning again toward childish treble, pipes
And whistles in his sound. Last scene of all,
That ends this strange eventful history,
Is second childishness and mere oblivion,
Sans teeth, sans eyes, sans taste, sans everything. [9]

Victor Hugo – French writer extraordinaire[10]

Victor-Marie Hugo, French writer – 1882 to 1885 - is described as poet, playwright, novelist, dramatist, essayist, visual artist, statesman, human rights campaigner, and perhaps the most influential exponent of the Romantic Movement in France. Some consider him as the most important of the French Romantic writers. Though regarded in France as one of that country's greatest poets, he is better known abroad for such novels as *Notre-Dame de Paris* (1831) and *Les Misérables* (1862). *Les Miserables*, first published in 1862, is considered one of the greatest novels of the 19th century. [10]

"In the English-speaking world, the novel is usually referred to by its original French title. However, several alternatives have been used, including The Miserables, The Wretched, The Miserable Ones, The Poor Ones, The Wretched Poor, The Victims and The Dispossessed. Beginning in 1815 and culminating in the 1832 June Rebellion in Paris, the novel follows the lives and interactions of several characters, particularly the struggles of ex-convict Jean Valjean and his experience of redemption. [10]

Examining the nature of law and grace, the novel elaborates upon the history of France, the architecture and urban design of Paris, politics, moral philosophy, antimonarchism, justice, religion, and the types and nature of romantic and familial love. Les Misérables has been popularized through numerous adaptations for film, television and the stage, including a musical."

Upton Sinclair, a brilliant author of more than a hundred books, described the novel as "one of the half-dozen greatest novels of the world", and remarked that Hugo set forth the purpose of Les Misérables in the Preface.

"So long as there shall exist, by reason of law and custom, a social condemnation, which, in the face of civilization, artificially creates hells on earth, and complicates a destiny that is divine with human fatality; so long as the three problems of the age—the degradation of man by poverty, the ruin of women by starvation, and the dwarfing of childhood by physical and spiritual night—are not solved; so long as, in certain regions, social asphyxia shall be possible; in other words, and from a yet more extended point of view, so long as ignorance and misery remain on earth, books like this cannot be useless." [10]

Towards the end of the novel, Hugo explains the work's overarching structure:

> "The book which the reader has before him at this moment is, from one end to the other, in its entirety and details ... a progress from evil to good, from injustice to justice, from

falsehood to truth, from night to day, from appetite to conscience, from corruption to life; from bestiality to duty, from hell to heaven, from nothingness to God. The starting point: matter, destination: the soul - the hydra at the beginning, the angel at the end."[10]

The novel contains various subplots, but the main thread is the story of ex-convict Jean Valjean, who becomes a force for good in the world but cannot escape his criminal past. The novel is divided into five volumes, each volume divided into several books, and subdivided into chapters, for a total of 48 books and 365 chapters. Each chapter is relatively short, commonly no longer than a few pages.

The novel as a whole is one of the longest ever written, with 655,478 words in the original French.

And as pointed out recently by my grandson, Jordan Smirl - Victor Hugo has been quoted as having stated: "The paradise of the rich is made out of the hell of the poor," - a sad, but true analogy of an economy controlled and enjoyed by the rich and powerful at the peril of the workers!

CHAPTER **2**

Other Well-known Voices

Ruth Bader Ginsburg

Between 1983 and 2019, Ruth Bader Ginsburg, Supreme Court Justice, wrote opinions on more than 200 of the more than 400 cases heard by the U.S. Supreme Court. [11]

Ruth Bader Ginsburg, the lady who plowed new ground and made her humanitarian mark on our history and our future, exemplifies the struggle for the rights of women and minorities.

A discussion of government for the common good should include the impact of Ruth Bader Ginsburg, a woman small in size but huge in stature. The Supreme Court Justice earned her law degree from Harvard, one of just nine women in her class more than 500 men. She went on to become the first female tenured professor at Columbia Law School. She founded the Women's Rights Project at the American Civil Liberties Union and helped to craft the legal case for women's equality, bringing cases before the Supreme Court where in 1993 she would become only the second female Supreme Court Justice in history.[11]

Ruth Bader Ginsburg fought on behalf of women who were discriminated against, people of color whose votes were suppressed, employees abused by their bosses, gay and lesbian people punished by bigots, and defendants denied due process of law and equal protection under the law.

OTHER WELL-KNOWN VOICES

"As the co-founder of the first law journal to focus on women's rights and as the general counsel for the Women's Rights Project of the American Civil Liberties Union, Ginsburg quickly proved herself a formidable opponent: she won five out of the six cases she argued before the Supreme Court between 1973 and 1976. She had a genius for the long game — taking on 'wedge' suits on men's rights to crack apart, brick by brick, the barriers that women faced." [11]

Quoting Ruth Bader Ginsburg:

"Real, enduring change, happens one step at a time."

"Fight for the things that you care about, but do it in a way that will lead others to join you."

Ginsburg had an uncommonly good relationship with Antonin Scalia while they were both Supreme Court Justices. When Scalia, a staunch conservative, was asked how he and Ginsburg, a liberal, who had opposing views, could have such a good friendship, Scalia was quoted as having said: "I attack ideas. I don't attack people. Some very good people have some very bad ideas. And if you can't separate the two you need to get another day job."

"Justice Ginsburg realized that she was but one member of a movement to give voice to the voiceless. Speaking of Pauli Murray and colleague Dorothy Kenyon at the American Civil Liberties Union, she proclaimed, 'We're standing on their shoulders. We're saying the same things they said, but now at last society is ready to listen.' As her words reveal, Justice Ginsburg understood that she was not here to win a marathon alone, but rather to run one leg in a relay race toward a more equitable future. She knew that no single person achieves lasting progress on the people's behalf—not even the Notorious RBG.

Indeed, Justice Ginsburg herself reflected both the aspirations and the limits of the feminism she espoused and ultimately came to embody for the nation, as her own hiring record for African American clerks (only one throughout her 38-year career as a judge) did not align with her progressive jurisprudence on racial issues and her personal embrace of Pauli Murray and Professor Kimberlé Crenshaw's theory of intersectionality, of which Murray's words about the "indivisibility of her identity" are a foremother.

As we reflect on Justice Ginsburg's legacy, we must guard against creating a caricature of the progressive icon she has become—the feisty, radical woman who single-handedly achieved constitutional protections for women and who represents all of our aspirations for a more just and equitable society that is truly of, by, and for the people. To do so, would be a disservice to the many invisible fighters along the way and, more so, to the Justice's truly impressive memory. As the president asserted, "Whether you agree with her or not, she was an amazing woman who led an amazing life."

Justice Ginsburg fully recognized that progress is incremental and often requires compromise. She is quoted as having said: "The Court generally moves in small steps rather than in one giant step, Real change, enduring change happens one step at a time."

Justice Ginsburg knew how to play the long game. She realized that one did not have to win every argument to continue the march toward justice, and she lost many battles along the way. "I do hope that some of my dissents will someday be the law," she said. "That's the dissenter's hope: that they are writing not for today but for tomorrow. Dissents speak to a future age." Fortunately, Justice Ginsburg lived to see her dissent in Ledbetter v. Goodyear Tire & Rubber Company, which she read with great passion directly from the bench—in my imagination, with her Rutgers Law School faculty experience at the back of her mind—become law through the Lily Ledbetter Fair Pay Act of 2009." [13]

Any review of the career of Ruth Bader Ginsburg should conclude that she was a stalwart in the struggle for equality in the use of power and privilege.

Woody Guthrie[14]

Woody Guthrie became a voice for equality of opportunity in 1940 with his penetrating songs of equal rights. He continued in his protest of power and privilege until his death in 1967. [14]

Woodrow Wilson Guthrie - July 14, 1912 – October 3, 1967, was an American singer-songwriter and one of the most significant figures in early American folk music. His work focused on themes of American socialism and anti-fascism. His music includes songs such as "This Land Is Your Land:" [14]

This land is your land

This land is your land, and this land is my land
From the California to the New York Island,
From the Redwood Forest, to the Gulf stream waters,
God blessed America for me.
This land was made for you and me.

As I went walking that ribbon of highway
And saw above me that endless skyway,
And saw below me the golden valley, I said:
God blessed America for me.
This land was made for you and me.

Guthrie wrote hundreds of country, folk, and children's songs. Many of his recorded songs are archived in the Library of Congress. Songwriters such as Bob Dylan, Johnny Cash, Bruce Springsteen, Pete Seeger, Jerry Garcia and many others have acknowledged Guthrie as a major influence on their work. He frequently performed with the slogan "This machine kills fascists" displayed on his guitar.

During his early teens, Guthrie learned folk and blues songs from his parents' friends. He married at 19, but with the advent of the dust storms that marked the Dust Bowl period, he left his wife and three children to join the thousands of Okies who were migrating to California looking for employment.

In February 1940 he wrote his most famous song, "This Land Is Your Land". He said it was a response to what he felt was the overplaying of Irving Berlin's "God Bless America" on the radio.

Woody, his mother and his two daughters all died of Huntington's disease. His son Arlo took up the torch and has become a popular voice singing some of the songs his father became famous with, along with others of his creation. [14]

CHAPTER **3**

Recent Common Good Voices

IN THE 21ST century, a time which we can readily relate to, the voices of common good include people such as song writers and performers Leonard Cohen, Arlo Guthrie, John Lennon, and Bob Marley, to mention a few.

These famous musicians' contributions are memorable and emotionally rewarding. Their meaning and value to our understanding of life's pleasures and challenges are endless and a treasure for all music lovers and listeners.

Those worthy of acknowledgement are too numerous to be all-inclusive, but a few are mentioned here as examples of the value in defining life's challenges and pleasures in poetic expression.

Leonard Cohen [15]

Leonard Cohen, who was with us from 1934 until 2016, was a Canadian singer-songwriter, poet, and novelist. His work explored religion, politics, depression, sexuality, loss, death and romantic relationships. His uncommon expression of such a broad spectrum of life's pleasures, divisions, trials and tribulations will surely survive any loss of our memory and appreciation of his contribution to the understanding and acceptance of emotional and lasting conditions of life which we all experience.

In support of the subject of this writing, special note of Cohen's contribution includes lines from his songs: "Everybody Knows," "Democracy," and "Anthem."

Cohen, as a great poet, songwriter, and performer - expresses life's disappointments, pursuits, and challenges. The ability to console and define us by expressing the pleasures, failures, and disappointments in life in such a meaningful manner inspires many of us to be Leonard Cohen fans.

Everybody Knows was released in 1988 by Cohen along with Sharon Robinson, a backup singer and songwriter, who provided verse and format to Cohen's original version. It has recently become one of his most recognized songs with personal meaning for most listeners. The song, loaded with powerful implications throughout the poetic verses, provides meaning by expressing life's maladies.

Everybody Knows[16]

> Everybody knows that the dice are loaded
> Everybody rolls with their fingers crossed
> Everybody knows that the war is over
> Everybody knows the good guys lost
> Everybody knows the fight was fixed
> The poor stay poor, the rich get rich
> That's how it goes
> Everybody knows

In his song, Democracy, which was released in 1992, Cohen saw the country as unprepared for the realities of the world we live in today. He expresses fears that our world is quickly being destroyed, and that we are guilty of allowing these atrocities to continue. It seems more of a warning, a message for humanity, than a song of ultimate peril.

RECENT COMMON GOOD VOICES

Democracy[17]

It's coming through a hole in the air
From those nights in Tiananmen Square
It's coming from the feel
That this ain't exactly real
Or it's real, but it ain't exactly there
From the war against disorder
From the sirens night and day
From the fires of the homeless
From the ashes of the gay
Democracy is coming to the USA

Anthem was also released in 1992, Cohen is quoted as having stated: "I think it is one of the best songs I have written, maybe the best," as told to music critic Robert Hilburn in 1995. "I knew that song was everything that my whole work and life had somehow gathered around. It is absolutely true to me." Rebecca De Mornay, Cohen's companion at the time, was credited as the song's co-producer.

Anthem[18]

The birds they sang at the break of day
Start again I heard them say
Don't dwell on what has passed away
Or what is yet to be

Ah, the wars they will be fought again
The holy dove, she will be caught again
Bought and sold, and bought again
The dove is never free ….

Cohen's contribution to music and poetry spanned several decades and touched the hearts of listeners with his soulful way of bringing the emotions to life. His poetic expression in song exemplify this subject matter by shining light on reality in simple and meaningful terms of which we can all relate in our own way.

Bob Marley

Bob Marley was a Jamaican singer, songwriter, and musician with us from 1945 until 1981. Considered one of the pioneers of reggae, his musical career was of a distinctive vocal and songwriting style.

In the 1981 Rolling Stone obituary, Bob Marley biographer Timothy White wrote, "The pervasive image of Bob Marley is that of a gleeful Rasta with a croissant-sized spliff clenched in his teeth, stoned silly and without a care in the world. But, in fact, he was a man with deep religious and political sentiments who rose from destitution to become one of the most influential music figures in the last 20 years."

Marley's stature and influence as a singer, songwriter, and international pop-culture prophet have only grown since those words were written. He is covered by countless singers, sampled and quoted by just as many hip-hop acts whose artistic DNA is shaped profoundly by the Jamaican music Marley defined. His social commitment remains an inspiration to many activists and his songs of freedom have become universal hymns.

"Marley sang about tyranny and anger, about brutality and apocalypse, in enticing tones, not dissonant ones," Mikal Gilmore wrote in 2005. "His melodies take up a resonance in our minds, in our lives, and that can provide admission to the songs' meanings. He was the master of mellifluent insurgency."

Get Up, Stand Up[19]

Get up, stand up. Stand up for your rights.
Get up, stand up. Stand up for your rights. Stand up for your rights.

Preacher man, don't tell me
Heaven is under the earth
I know you don't know
What life is really worth
It's not all that glitters is gold
Half the story has never been told
So now you see the light,
Stand up for your rights

Redemption Song[20]

Old pirates, yes, they rob I
Sold I to the merchant ships
Minutes after they took I
From the bottomless pit

But my hand was made strong
By the hand of the almighty
We forward in this generation
Triumphantly

Won't you help to sing
These songs of freedom
'Cause all I ever have
Redemption songs
Redemption songs

Bob Marley left us in 1981 at age 36 due to the spread of melanoma to his lungs and brain. His final words to his son Ziggy were "Money can't buy life." That one should sink in for all of us. His early loss of life is a loss for all of us. He would surely have added many more songs of protest and lessons for life in general for decades to come.

FOOTPRINTS

John Lennon[21]

John Lennon, who was with us from 1940 to 1980, was a well-known English musician who gained prominence as a member of the Beatles. He recorded many songs with Paul McCartney, his wife Yoko Ono, and individually.

Lennon recorded over 150 songs as a solo artist. Between 1968 and 1969, Lennon released three albums with wife Yoko Ono, as well as two singles, "Give Peace a Chance" and "Cold Turkey. His debut single before the Beatles' break-up was "Instant Karma!"

Imagine, was released in 1971 with Phil Spector and featured appearances by George Harrison. Imagine features songs with lyrics discussing peace and love.

Its title track, "Imagine," is regarded as one of Lennon's finest songs.

Songs by Lennon and Ono featured lyrics discussing political and social issues and topics such as sexism, incarceration, colonialism and racism.

Mind Games of 1973, Lennon's first self-produced album, included love songs, hard rockers, and bouts of humor. Walls and Bridges 1974, recorded during his 18-month separation from Ono, features rock and pop songs that reflected Lennon's feelings at the time. Rock 'n' Roll, in 1975, was an album of late 1950s and early 1960s rock songs, included songs such as "Stand by Me", "Peggy Sue" and "You Can't Catch Me". After Rock 'n' Roll, Lennon took a five-year hiatus from the music industry to raise his son Sean, aside from occasional demos.

Lennon returned to music in 1980 with Ono on the album Double Fantasy. Co-produced by Jack Douglas, the album's songs primarily focus on the couple's relationship, emphasizing their love for each other and their son, Sean, with some songs discussing Lennon's hiatus.

Lennon was murdered in 1980 outside his apartment in New York City. In the years following his death, many previously unissued songs have seen release on other albums, including Milk and Honey (1984), Menlove Ave. (1986), and Anthology (1998). In 2020, to celebrate what would have been Lennon's 80th birthday, Ono and his son, Sean, released the box set, Gimme Some Truth. The Ultimate

Mixes contained newly remixed versions of 36 of Lennon's songs.[21]
One of his most memorable contributions is, of course, "Imagine."

Imagine[22]

Imagine there's no heaven
It's easy if you try
No hell below us
Above us only sky
Imagine all the people living for today

Imagine there's no countries
It isn't hard to do
Nothing to kill or die for
And no religion too
Imagine all the people living life in peace, you

You may say I'm a dreamer
But I'm not the only one
I hope someday you'll join us
And the world will be as one

Imagine no possessions
I wonder if you can
No need for greed or hunger
A brotherhood of man
Imagine all the people sharing all the world

You may say I'm a dreamer
But I'm not the only one
I hope someday you'll join us
And the world will be as one.

I don't know who ever said it any better than John Lennon?

Arlo Guthrie

Arlo Davy Guthrie is an American folk singer-songwriter. He is known for singing songs of protest against social injustice, and storytelling while performing songs, in following in the tradition of his father, Woody Guthrie.

Some of his more popular songs include "Alice's Restaurant" and "City of New Orleans."

Alice's Restaurant[23]
This song is called Alice's Restaurant
And it's about Alice, and the Restaurant,
But Alice's Restaurant is not the name of the restaurant
That's just the name of the song
And that's why I called the song Alice's Restaurant.
You can get anything you want at Alice's Restaurant
You can get anything you want at Alice's Restaurant
Walk right in, it's around the back
Just a half a mile from the railroad track
You can get anything you want at Alice's Restaurant

City of New Orleans[23]
Riding on the City of New Orleans
Illinois Central Monday morning rail
Fifteen cars and fifteen restless riders
Three conductors and twenty-five sacks of mail
All along the southbound odyssey
The train pulls out at Kankakee
Rolls along past houses, farms and fields
Passin' trains that have no name
Freight yards full of old black men
And the graveyards of the rusted automobiles
Good morning America how are you?
Say, don't you know me? I'm your native son
I'm the train they call the City of New Orleans
I'll be gone five hundred miles when the day is done.

Burt Bacharach

Burt Bacharach, born 1928 in Kansas City, Missouri, brought the subject of love and fidelity to light from the 1950s through the 1980s with his many popular songs. His simple but poignant poetry in his song Alfie, a song about realization of what really matters in lasting relationships, is one we all know and which has been famous with other performers and movie producers:

Alfie[24]
By Burt Bacharach

What's it all about Alfie
Is it just for the moment we live?

What's it all about
When you sort it out, Alfie
Are we meant to take more than we give
Or are we meant to be kind?

And if, if only fools are kind, Alfie
Then I guess it is wise to be cruel
And if life belongs only to the strong, Alfie
What will you lend on an old golden rule?

As sure as I believe there's a heaven above
Alfie, I know there's something much more
Something even non-believers can believe in

I believe in love, Alfie
Without true love we just exist, Alfie
Until you find the love you've missed
You're nothing, Alfie

When you walk let your heart lead the way
And you'll find love any day Alfie, Alfie

CHAPTER 4

More Voices for the Common Good

EXAMPLES OF THE exercise and abuse of power and privilege seem endless when reviewing our many great contributions to the subject by American and other authors. Some of those deemed pertinent are included here.

Harriet Beecher Stowe – Author[25]

Harriet Elisabeth Beecher Stowe, who was with us from 1811 until 1896, was an American author and abolitionist who became best known for her novel, Uncle Tom's Cabin, a widely read classic which depicts the harsh conditions experienced by enslaved African Americans. The novel had a profound effect on attitudes toward African Americans and slavery in the U.S., and is said to have helped lay the groundwork for the American Civil War.

Uncle Tom's Cabin was the best-selling novel and the second best-selling book of the 19th century, following the Bible.[42]

Harriet Beecher Stowe's novel focused considerable light on the tragic effects bestowed upon blacks of the time. The book was both acclaimed and criticized by numerous scholars soon after publication and periodically for decades thereafter, both by black authors and the other scholars nationally and internationally. Nevertheless, it remains a classic among U.S. scholastic institutions and continues to be widely read.

For this publication, Stowe's writing displays one of the saddest examples of the abuse of power and privilege in our history - slavery.

Anna Elizabeth Dickinson[26]

Anna Dickinson, an American lecturer and abolitionist, was with us from 1842 until 1932. She began publishing at the age of 14 and began public speaking on "Women's Rights and Wrongs." Much of her work was in behalf of the Republican Party. She was known for her fiery oratory and passionate delivery. She was the first woman to give a political address before the United States Congress.

Dickinson's subjects titled as: "Woman's Work and Wages," "White Sepulchres," "Demagogues and Workingmen," (which is what she called Mormons), and "A Ragged Register (of People, Places, and Opinions), a memoir.

Her message always focused on the excesses and denials of power and privilege.

Robert Reich – Professor and Writer[27]

Robert Reich is a true stalwart of common good principles and actions and a leading advocate for equality of opportunity, vehemently opposing capitalistic inequality.

Robert Reich's latest book is: "The System: Who Rigged It, How to Fix It." He is Chancellor's Professor of Public Policy at the University of California at Berkeley and Senior Fellow at the Blum Center. He served as Secretary of Labor in the Clinton administration, for which Time Magazine named him one of the 10 most effective cabinet secretaries of the twentieth century. He has written 17 other books, including the best sellers "Aftershock," "The Work of Nations," "Beyond Outrage," and "The Common Good." He is a founding editor of the American Prospect magazine, founder of Inequality Media, a member of the American Academy of Arts and Sciences, and co-creator of the award-winning documentaries "Inequality For All," streaming on YouTube, and "Saving Capitalism," now streaming on Netflix.

As stated in introducing his book "The Common Good," Reich demonstrates the existence of a common good, and argues that it is this that defines a society or a nation. Societies and nations undergo virtuous cycles that reinforce and build the common good, as well as vicious cycles that undermine it. Over the course of the past five decades, Reich contends, America has been in a slowly accelerating vicious cycle--one that can and must be reversed. But first we need to weigh what really matters, and how we as a country should relate to honor, shame, patriotism, truth, and the meaning of leadership.

Noam Chomsky – Linguist and Philosopher[28]

Most of us know the name of Noam Chomsky, son of a Hebrew scholar, an American linguist, philosopher, cognitive scientist, historian, social critic, and political activist. Sometimes called "the father of modern linguistics," Chomsky is also a major figure in analytic philosophy and one of the founders of the field of cognitive science. He is said to be the most cited living author and the most important social critic in the world. He became a full professor teaching philosophy and linguistics at MIT at age 32, has authored many books and has received some 37 honorary degrees.

Chomsky liked to quote a famous unnamed Brazilian general of about 1970 who stated about the economy: "… the economy is doing fine – it's just the people that aren't" – a picture worth a thousand words.

Chomsky speaks of Aristotle's belief that democracy should be fully participatory to achieve lasting prosperity for everyone and that if you have extremes of poor and rich, you can't really have democracy. He felt that Aristotle's belief was that to maintain democracy we should reduce poverty.

He speaks of James Madison who believed the solution was to reduce democracy, which is expressed by John Jay who believed that the people who own the country should govern it. He designed a

MORE VOICES FOR THE COMMON GOOD

system which assured that democracy could not function.

Throughout our history, political power has been in the hands of those who own the country. Regardless of legislative struggles for the common good, the power is in the hands of those with the financial wherewithal to maintain political power, the rich and powerful.

"Modern science and technology can relieve men of the necessity for specialized, imbecile labor. They may, in principle, provide the basis for a rational social order based on free association and democratic control if we have the will to create it."

"A basic principle of modern state capitalism is that costs and risks are socialized to the extent possible, while profit is privatized. Noam Chomsky"

"It's an unjust system taken to toxic levels and the logical end results are becoming apparent everyday - more unemployment, homelessness, poverty, crime, environmental destruction, medical and food insecurity and basically misery."

Chomsky's states that speaking truth to power makes no sense because they already know the truth, It would be better to speak to the powerless and maybe they will act to dismantle illegitimate power. He believes that "if you look at what is happening it is not very pretty and if you extrapolate that into the future it is very ugly. Nobody ever said it was going to be easy." (How the World Works page 314)

Other statements by Chomsky:

"Investment is supposed to be as risk-free as possible. No corporation wants free markets – what they want is power."

"Much of disparity between blacks and whites is actually a class difference, and the gap between poor and rich is also enormous. But you're not allowed to talk about class in the US. "

"What the public wants is called "politically unrealistic." Translated into English that means power and privilege are opposed to it."

"This is a business-run, huckster society and its primary value is deceit."

Comments of Noam Chomsky from over the decades:

Crime: in the suites vs. in the streets

"The media pays a lot of attention to crime in the streets, which the FBI estimates cost about $4 billion a year. The Multinational Monitor estimates that white-collar crime- what Ralph Nader calls "crime in the suites"-costs about $200 billion a year. That generally gets ignored.

Although crime in the US is high by the standards of comparable societies, there's only one major domain in which it's really off the map- murders with guns. But that's because of the gun culture.

Drug related crimes, usually pretty trivial ones, are mostly what are filling up the prisons. I haven't seen many bankers or executives of chemical corporations in prison. People in the rich suburbs commit plenty of crimes, but they're not going to prison at anything like the rate of the poor.

The police obviously find it much easier to make an arrest on the streets of a black ghetto than in a white suburb. A very high percentage of incarceration is drug-related, and it mostly targets little guys, somebody caught peddling dope."[28]

Thomas L. Friedman – Columnist and Writer[29]

Thomas Friedman has been awarded the Pulitzer Prize three times for his work at the New York Times where he serves as a foreign affairs columnist. His publication - "Hot, Flat, and, Crowded," provides us with revealing information about the current need for a "green revolution," as he calls it, and how it can renew America.

To quote the relevant statements, facts, and figures in Friedman's book would require quoting the book in its entirety. They are too numerous to mention them all but no discussion of the subject would be complete without considerable inclusion.

Friedman tells us that we cannot stop global warming because too much CO_2 is already baked into our future, but that we can reduce the rate of global warming and reduce the odds of a disaster. He tells us that all we can do is try to fit in as species and if the species doesn't fit in with Mother Nature it gets kicked out – and that now every time you look in a mirror you are looking at an endangered species.

He tells us that we are now at a point where business as usual is expected to produce as much as an eleven degree Fahrenheit in average global temperature by 2100. He quotes Al Gore: "That is so unthinkable it would bring civilization to a screeching halt and tear apart the fabric of life."

He tells us that Ted Turner said: "there are too many people using too much stuff."

Friedman tells us that he can't think of anything more stupid than us buying as much stuff as we can from the people who hate us the most.

He tells us that "our addiction to oil makes global warming warmer, petro-dictators stronger, clean air dirtier, poor people poorer, democratic countries weaker and radical terrorists richer."

Friedman quotes Jared Diamond: "the world is already running out of resources and it will do so even sooner if China achieves the American level of consumption rates."[29]

Paul Krugman – Professor and Economist[30]

Paul Krugman, a confessed liberal, Distinguished Professor of Economics at the Graduate Center of the City University of New York, columnist for The New York Times, and former professor of economics at Princeton University, addressed the subject at length in his 2007 book: "The Conscious of a Liberal." His statement at the time: "liberals are always trying to enfranchise citizens, while conservatives are always trying to block citizens from voting" He further stated: "I believe in a relatively equal society, supported by institutions that limit extremes of wealth and poverty. I believe in democracy, civil liberties, and the rule of law. That is what makes me a liberal."

Staying on the case, Krugman posted a follow up article: "Tax the Rich, Help America's Children." He states that "there is overwhelming evidence that helping children, in addition to being the right to do, has big economic benefits. Food stamps make them healthier, more productive adults. Children with preschool are more likely to graduate from high school and go to college, and that investing in children is more beneficial than investing in physical infrastructure."

Paul Krugman provides a profound voice for liberalism in the present time. His writing will hopefully continue to carry the torch of equality of opportunity, the ultimate goal of liberalism, until other voices join the chorus of equal rights to counter the domination of economic opportunity by the rich and powerful. It becomes more obvious each year that we all share the limited resources of the planet but that many those with power and privilege are so consumed by greed that they ignore the ultimate demise for future generations.

Krugman tells that we have had progressive taxation since 1916. It is nothing new. In 1905 Theodore Roosevelt it was essential to prevent the "inheritance or transmission of fortunes swollen beyond healthy limits," and that he called for heavy progressive taxation on estates as well. [30]

Thomas Sowell – Professor and Writer[31]

Thomas Sowell of The Hoover Institution, Stanford University is the author of Wealth, Poverty and Politics 2016. He takes his stand as he quotes David S. Landes from "The Wealth and Poverty of Nations:" "The world has never been a level playing field, and everything costs."-

In "Wealth, Poverty and Politics:" (Goals – page 407) Sowell tells us that "Monstrously appalling things done by some peoples to others darken the history of every region of the planet, but descendants of peoples guilty of the worst or most extensive villainies of the past are by no means always the most prosperous peoples today. Conversely, few peoples have been persecuted for so many centuries, in so many parts of the world as the Jews, who today prosper and achieve.

No explanation of glaring economic differences by geography, demography, culture, or other impersonal factors has ever enjoyed the sudden worldwide acceptance and devotion as Marxism had in the twentieth century – a theory, belief system and agenda based on the assumption that the poor are poor because the rich exploit them.

Yet none of the Communist countries established around the world ever achieved a standard of living for ordinary people equal to that in a number of capitalist economies."

"In a world where we are all beneficiaries of enormous windfall gains that our forebears never had, are we to tear apart the society that created all this, because some people's windfall gains are greater of less than some other people's windfall gains?" [31]

Peter F. Drucker – Author and Educator[32]

Austrian-American management consultant, educator, and author, is often quoted about the modern business world. His book, "The Age of Discontinuity" of 1968, offers challenges that existed in 1968 but still linger today.

Some of his statements are worthy of note today:

"The growth for education and training will be in continuing adult education. Online delivery is the trigger for this growth, but the demand for lifetime education stems from profound changes in society. We live in an economy where knowledge, not buildings and machinery, is the chief resource and where knowledge-workers make up the biggest part of the work force."

"Trying to predict the future is like trying to drive down a country road at night with no lights while looking out the back window."

"No one needs to tell us that our age in an age of infinite peril. No one needs to be told that the central question we face with respect to man's future is not what it shall be, but whether it shall be."[32]

Albert Gore – Former Vice President of the United States[33]

Al Gore, son of Albert Gore, Sr. former U.S. Senator, is the person who some believe was elected president of the United State in the year 2000, although the disputed outcome was cut short of resolution by the right-leaning Supreme Court of the United States. Gore's has been an incessant voice for the common good for decades. He served as Vice President under Bill Clinton from 1993 to 2001.

Gore, a prolific writer, has authored several books supporting the imperativeness of the principle of common good in our governance. His literary masterpieces include: "Earth in the Balance," "An Inconvenient Truth," "Joined at the Heart," "The Spirit of Family," "Common Sense Government," "Businesslike Government," and "The Assault on Reason."

His latest book, "The Assault on Reason," was published in 2007. In introducing his message he quotes Robert Byrd of West Virginia who said on the Senate floor speaking about the present war: "This Chamber is silent, ominously, dreadfully silent. We stay passively mute in the United States Senate."

Gore describes a "Politics of Wealth," whereby wealth and power have become concentrated in the hands of a few who consolidate and perpetuate their control at the expense of the many. He quotes the Roman historian Plutarch who was quoted as having stated that: "an imbalance between the rich and the poor is the oldest and most fatal ailment of all Republics." (Plutarch lived from AD 46 until AD 119 – more than 2000 years ago. Some things never change when it comes to the distribution of power and privilege.)

Gore quotes Adam Smith, the founder of capitalism, as having stated: "All for our-selves and nothing for other people, in every age of the world, has been the maxim of the masters of mankind."

He quotes Immanuel Kant, the most influential European philosopher of the Enlightenment: "The enjoyment of power inevitably corrupts the judgement of reason, and perverts its liberty,"

And he quotes Upton Sinclair as having said, "It is difficult to get a man to understand something when his salary depends upon his not understanding it."

Gore states that: "Greed and wealth now allocate power in our society, and that power is used in turn to further increase and concentrate wealth and power in the hands of the few." He believes that we must concentrate on re-empowering the people of the United States with the ability and the inclination to fully and vigorously participate in the national conversation of democracy.

Gore, in his ending comment in "The Assault on Reason," poses the question: "Will we continue to live as a people under the rule of law as embodied in our Constitution, or will we fail future generations by leaving them a Constitution far diminished from the charter of liberty we have inherited from our forebears?"[33]

William Greider, author of "The Soul of Capitalism"[34]

William Greider, in his comments in the year 2003, tells us that most citizens are pretty much withdrawn from the arena of political action. He expects that most Americans are aware of the deleterious impact of capitalism but they feel powerless to do anything about it.

"We can observe the injustices of "limited liability" in the laborious legal cleanups following the debacles of Enron, Worldcom, Global Crossing and many other bankruptcies. The insiders held onto their mansions and personal fortunes. The losers-creditors, suppliers, shareholders, employees-fought over the scraps, with their claims stacked in descending order of priority-bankers first, employees last. Limited liability dilutes the ideal of personal responsibility for one's actions.

The corporation functions as a principal source of American inequality, concentrating both power and wealth at the top. Greider quotes Charles Perrow in "Organizing America": "The accumulation of wealth and power through large organizations is the modern device for generating inequality." Greider states that "… the corporation functions as a principal source of American inequality, concentrating power and wealth at the top."

He tells us that government has been a central actor in developing American capitalism and that capitalism would be bereft and feeble if government were not always standing close by with a helping hand. They provide public capital to finance innovative ventures regarded as too risky for private investors and that the libertarian vision in which enterprise prospers free of political interference in a myth.

Greider discusses the way that various states and communities are played against each other when corporations are seeking areas to attract workers to fill their demands, extracting large benefits of capital assets and tax breaks as an attraction to locate in the area under the expectation of economic advantage. This can be defined as actual capital investment in the given corporation which requires no repayment. This is actually unredeemable investment by the taxpayers while the given corporation makes generous political donations over time to the political entity. "You scratch my back and I'll scratch yours."[34]

Upton Sinclair, author of "The Jungle" – 1906 [35]

Before we rest our case about the effects of capitalism on society we should go back to the earlier days of writings about the abuses of the workers for the pleasure of the capitalists. Upton Sinclair introduced this to the readers more than a hundred years ago with his publishing of "The Jungle." This book was about the early days of the meat packing industry and the physical and mental abuses which were imposed upon the workers in the name of profit for the industry.

Sinclair described the site of Chicago's meatpacking industry as "the greatest aggregation of labor and capital ever gathered in one place." He revealed that the supreme achievement of American capitalism was its greatest disgrace. It was a revelation of the sale of meat through pervasive wage and price fixing and the unrelenting exploitation of the stockyard workforce which President Theodore Roosevelt called it "muckraking."

Upton Sinclair was a precocious child who entered college at the age of thirteen. He wrote short stories early in his writing career but came into his own as a socialist thinker in the early 1900s and participated with socialist groups with such socialists as Jack London and Clarence Darrow. He is known as a significant contributor to American social thought focusing on 'wage slavery.'". (Preface to The Jungle 2003 publication)

Other writings of those times regarding worker exploitation and the ties between government and business would include: "Wrath against Commonwealth" by Henry Lloyd, "History of the Standard

Oil Company" by Ida Tarbell, and "The Shame of the Cities" by Lincoln Steffens

The above comments are from the introduction by Maura Spiegel of Columbia University in the 2003 publication of "The Jungle."[35]

Joseph E. Stiglitz – author of "Freefall" 2010 [36]

Joseph Stiglitz talks about the Great Recession that began in 2007 and which caused retirement and education funds of individuals to dwindle in value, a crisis beginning in America and then spreading globally when tens of millions lost their jobs and fell into poverty. He discusses the crisis as Wall Street destroyed homes, education, and jobs; while bankers fell all over themselves and sought government bailouts while resisting regulation to stave off future crises.

Stiglitz discusses devastation that the financial institutions brought to the economy, the unrelenting pursuit of profits and the increase of the pursuit of self-interest creating a moral deficit in finance, the incompetence and deception, risky subprime mortgages, financial leverage – all resulting in the exploitation of the poor and middle-class Americans.

There were no limits to acceptable behavior. "These episodes were marked by moral scruples that should make us blush, that a few of the most egregious personalities were marched off to jail, such as: Charles Keating, Michael Milken, Kenneth Lay, and Bernard Ebbers, and he added the name of Angelo Mozilo of Countrywide Financial to the list of the morally challenged. And a disparaging fact remains, that even after paying staggering fines, there were hundreds of millions of dollars left in their accounts."[36] (This seems to support the theory that crime pays)

Charles H. Ferguson – author of "Predator Nation" 2012[37]

Charles Ferguson tells us that reason he wrote this book is that "although there have been many books written about the financial crisis, the bad guys got away with it." He quotes the gangster Al Capone, who once said, "You can get much further in life with a kind word and a gun than with a kind word alone."

He tells us that "the flip side of the growth in American inequality is an obscene, morally indefensible decline in the fairness of American society – in education, job opportunities, income, wealth, health, and life expectancy. And that with exception of wealthy families, children in America are now less educated than their parents, and will earn less money than their parents and that the United States is transforming itself into one of the most unfair, most rigid, and least socially mobile of the industrialized countries."

Among Ferguson's "What Should be Done" section he hits on one that supports a main focus of this writing: "Control the impact of money on American politics. Lobbying and political contribution should be heavily taxed. Public sector and regulator salaries should be raised sharply, and some form of public campaign financing should be made mandatory."[37]

CHAPTER 5

Entertainers Opinions

Oliver Stone [38]

Oliver Stone, known as a controversial and politically-charged film producer/director, made some of the most politically enduring films of his time, including: "Platoon," "Wall Street," "Born on the Fourth of July," and "J.F.K.". He presented his take on a turbulent time wrought with sex, drugs and an endless war. He made a string of exceptional films that supported his reputation as one of Hollywood's most unique and compelling directors. He was a man uniquely concerned with the destructive actions of the country he loved. He made other films including: "Natural Born Killers,", "U-Turn," and "Alexander" which supported his place in cinematic history.

Stone was nominated for film awards in 1990, 1979, 1987, 1992, and 1996 and was awarded in 1990 as Best Director for "Platoon" and in 1992 as Best Director for "Born on the Fourth of July." Overall in his career he won more than 60 awards in various film festivals.

"Platoon" brought the true horror of the Vietnam War to the big screen. Based on Stone's own experiences as a soldier in the conflict, the film captivated millions of viewers all over the world. (Platoon – Wikipedia)

"Wall Street" is a 1987 American drama film, directed and co-written by Oliver Stone, which stars Michael Douglas, Charlie Sheen, Daryl Hannah, and Martin Sheen. The film tells the story of Bud Fox

(Charlie Sheen), a young stockbroker who becomes involved with Gordon Gekko (Michael Douglas), a wealthy, unscrupulous corporate raider.

"Born on the Fourth of July" is a 1989 American biographical anti-war drama film based on the eponymous 1976 autobiography by Ron Kovic. Directed by Oliver Stone, and written by Stone and Kovic, it stars Tom Cruise, Kyra Sedgwick, Raymond J. Barry, Jerry Levine, Frank Whaley, and Willem Dafoe. The film depicts the life of Kovic (Tom Cruise) over a 20-year period, detailing his childhood, his military service and paralysis during the Vietnam War, and his transition to anti-war activism. It is the second installment in Stone's trilogy of films about the Vietnam War, following Platoon (1986) and preceding Heaven & Earth (1993).

JFK is a 1991 American epic political thriller film that examines the events leading to the assassination of John F. Kennedy in 1963 and alleged cover-up through the eyes of former New Orleans district attorney Jim Garrison. Garrison filed charges against New Orleans businessman Clay Shaw for his alleged participation in a conspiracy to assassinate Kennedy, for which Lee Harvey Oswald was found responsible by the Warren Commission.

The film was directed by Oliver Stone, adapted by Stone and Zachary Sklar from the books "On the Trail of the Assassins" (1988) by Jim Garrison and Crossfire: "The Plot That Killed Kennedy" (1989) by Jim Marrs. Stone described this account as a "counter-myth" to the Warren Commission's "fictional myth."

Al Franken [39]

Al Franken makes the case very succinctly in his recent article in Rolling Stone: "Tax the Rich! Also the Very Affluent! But Mainly the Rich!"

His subtitle: "Republicans say tax cuts pay for themselves. They never do. How about we try something that actually does work?"

Al Franken, punished professionally for a perceived sexual degradation of women while doing what he has always done, being a

ENTERTAINERS OPINIONS

comedian, also has a very serious side. He expresses taxation based upon ability to pay in his Rolling Stone article in very meaningful terms.

Franken tells us that "taxing the rich is a good idea according to everyone who understands that trickle-down economics has failed spectacularly for decades. Yet, Republicans have weakened to estate tax so that $11.2 million of inheritance is exempt instead of the previous $5.5 million." He supports Biden's proposal to raise $80 billion from rich people like Trump who paid $750 in federal income taxes in 2017 and zero for many years prior. He states that "Biden hopes to raise $1.5 trillion over the next decade by cutting tax loopholes and raising the top individual tax rate to 39.6 percent, where it was when George W. Bush took office."

Franken states that "the GOP is barely a political party now – dedicated to nothing other than propping up the dumbest lie – that somehow the election was stolen from the malevolent, vindictive narcissist who got 7 million fewer votes than the other guy."

"In the wake of the Democrat's win in November, Republicans introduced 361 bills in state legislatures that would make it harder for people to access the ballot box."

"The fact is that every bit of what President Biden proposed is in everyone's best interest. If a bridge collapses, a Mercedes drops just as fast as a Hyundai."[39]

Steven Spielberg [40]

Schindler's List, a Steven Spielberg film, considered a cinematic masterpiece, has become one of the most honored films of all time. The film won seven Academy Awards, including Best Picture and Best Director. It also won every major Best Picture award and an exceptional number of additional honors. Among them were seven British Academy Awards; the Best Picture Awards from the New York Film Critics Circle, the National Society of Film Critics, the National Board of Review, the Producers Guild, the Los Angeles Film Critics, the Chicago, Boston and Dallas Film Critics; a Christopher Award;

and the Hollywood Foreign Press Association Golden Globe Awards.

The film presents the indelible true story of the enigmatic Oskar Schindler, a member of the Nazi party, womanizer, and war profiteer who saved the lives of more than 1,100 Jews during the Holocaust. It is the triumph of one man who made a difference, and the drama of those who survived one of the darkest chapters in human history because of what he did.[40]

Voices of the Music World

Musicians exemplify the struggle for equality of opportunity in poetic verse, which for many of us is more meaningful and more penetrating than spoken or written words. The contributions are innumerable in this musical expression of hope, misfortune, anguish, struggle, and accomplishment in all levels and walks of life. Some of our most well=known contributors are especially worthy of mention in a discussion of those who have had a positive impact on our quest for the common good.

Bob Dylan [41]

Bob Dylan, the great American singer-songwriter, author and visual artist, is often regarded as one of the greatest songwriters of all time. Whoever wrote more poetry to music than Bob Dylan? Dylan has been a major figure in popular culture during a career spanning nearly 60 years and is reported to have written more than 500 songs. Much of his most celebrated work dates from the 1960s, when songs such as "Blowin' in the Wind" (1963) and "The Times They Are a-Changin'" (1964) became anthems for the civil rights and anti-war movements. His lyrics during this period incorporated a range of political, social, philosophical, and literary influences, and appealing to the counterculture. Times were changing.

"Like a Rolling Stone":[41]

"Once upon a time you dressed so fine
Threw the bums a dime in your prime, didn't you?
People call say 'beware doll, you're bound to fall'
You thought they were all kidding you
You used to laugh about
Everybody that was hanging out
Now you don't talk so loud
Now you don't seem so proud
About having to be scrounging your next meal"

"How does it feel, how does it feel?
To be without a home
Like a complete unknown, like a rolling stone"

"The Times They Are A-Changin'"[41]

Come gather 'round people
Wherever you roam
And admit that the waters
Around you have grown
And accept it that soon
You'll be drenched to the bone
If your time to you is worth savin'
And you better start swimmin'
Or you'll sink like a stone
For the times they are a-changin'

The line it is drawn
The curse it is cast
The slow one now
Will later be fast
As the present now

FOOTPRINTS

Will later be past
The order is rapidly fadin'
And the first one now
Will later be last

"Blowin' In the Wind"[41]

Song by Bob Dylan, Tom Petty, and the Heartbreakers

"How many roads must a man walk down
Before you call him a man?
How many seas must a white dove sail
Before she sleeps in the sand?
Yes, and how many times must the cannonballs fly
Before they're forever banned?

The answer, my friend, is blowin' in the wind
The answer is blowin' in the wind

And how many ears must one man have
Before he can hear people cry?
Yes, and how many deaths will it take 'til he knows
That too many people have died?

The answer, my friend, is blowin' in the wind
The answer is blowin' in the wind"

Paul Simon – Simon & Garfunkel [42]

"Bridge Over Troubled Water"[42]

"When you're weary
Feeling small

When tears are in your eyes
I'll dry them all
I'm on your side
Oh, when times get rough
And friends just can't be found

Like a bridge over troubled water
I will lay me down

When you're down and out
When you're on the street
When evening falls so hard
I will comfort you
I'll take your part

Oh, when darkness comes
And pain is all around

Like a bridge over troubled water
I will lay me down"

Neil Young [43]

"After the Gold Rush"[43]

"Well, I dreamed I saw the knights in armor coming
Sayin' something about a queen
There were peasants singin' and drummers drumming
And the archer split the tree

There was a fanfare blowin' to the sun
That floated on the breeze
Look at mother nature on the run in the nineteen seventies
Look at mother nature on the run in the nineteen seventies

FOOTPRINTS

I was lyin' in a burned out basement
With a full moon in my eyes
I was hopin' for a replacement
When the sun burst through the skies

There was a band playin' in my head
And I felt like getting high
Thinkin' about what a friend had said
I was hopin' it was a lie

Well, I dreamed I saw the silver spaceships flying
In the yellow haze of the sun
There were children crying and colors flying
All around the chosen ones

All in a dream, all in a dream
The loading had begun
Flyin' mother nature's silver seed
To a new home in the sun"

Arlo Guthrie [44]

"Patriot's Dream"[44]

"Living now here but for fortune
Placed by fate's mysterious schemes
Who'd believe that we're the ones asked
To try to rekindle the patriot's dreams

Arise sweet destiny, time runs short
All of your patience has heard their retort
Hear us now for alone we can't seem
To try to rekindle the patriot's dreams

Can you hear the words being whispered
All along the American stream
Tyrants freed the just are imprisoned
Try to rekindle the patriot's dreams

Ah but perhaps too much is being asked of too few
You and your children with nothing to do
Hear us now for alone we can't seem
To try to rekindle the patriot's dreams"

Billy Joel [45]

"Piano Man"[45]

"It's nine o'clock on a Saturday
Regular crowd shuffles in
There's an old man sittin' next to me
Makin' love to his tonic and gin
He says: "Son can you play me a memory?"
I'm not really sure how it goes
But it's sad and it's sweet and I knew it complete
When I wore a younger man's clothes

Sing us a song you're the piano man
Sing us a song tonight
Well we're all in the mood for a melody
And you've got us feelin' alright

Now John at the bar is a friend of mine
He gets me my drinks for free
And he's quick with a joke or to light up your smoke
But there's someplace that he'd rather be

FOOTPRINTS

> He says Bill I believe this is killing me
> As a smile ran away from his face
> Well I'm sure that I could be a movie star
> If I could get out of this place"

Willie Nelson [46]

"Funny how Time Slips Away" [46]

Well, hello there
My, it's been a long, long time
How am I doin'?
Oh, well, I guess I'm doin' fine
It's been so long now and it seems that
It was only yesterday

Mmm, ain't it funny how time slips away?

How's your new love?
I hope that he's doin' fine
Heard you told him, yes, baby
That you'd love him till the end of time

Well, you know, that's the same thing
That you told me
Well, it seems like just the other day
Mmm, ain't it funny how time slips away

The Beatles [47]

"Let It Be" [47]

When I find myself in times of trouble, Mother Mary comes to me
Speaking words of wisdom, let it be
And in my hour of darkness she is standing right in front of me
Speaking words of wisdom, let it be

Whisper words of wisdom, let it be

And when the broken hearted people living in the world agree
There will be an answer, let it be
For though they may be parted, there is still a chance that they will see
Whisper words of wisdom, let it be

And when the night is cloudy there is still a light that shines on me
Shinin' until tomorrow, let it be
I wake up to the sound of music, Mother Mary comes to me
Speaking words of wisdom, let it be

Johnny Cash [48]

"Folsom Prison Blues" [48]

"I hear the train a-comin'; it's rollin' 'round the bend,

And I ain't seen the sunshine shine since I don't know when.

I'm stuck in Folsom Prison and time keeps draggin' on,
But that train keeps a-rollin'
On down to San Antone.

FOOTPRINTS

"bet there's rich folks eatin' in a fancy fining car.
They're prob'ly drinkin' coffee and smokin' big cigars,
But I know I had it comin', I know I can't be free,
But those people keep a-movin',
And that's what tortures me.
Well, if they freed me from this
Prison, if that railroad train was mine,
I bet I'd move on over a little farther down the line,"

Stevie Wonder [49]

"Higher Ground"[49]

"People keep on learnin'
Soldiers keep on warrin'
World keep on turnin'
Cause it won't be too long

Powers keep on lyin'
While your people keep on dyin'
World keep on turnin'
Cause it won't be too long

I'm so darn glad he let me try it again
Cause my last time on earth I lived a whole world of sin
I'm so glad that I know more than I knew then
Gonna keep on tryin'
Till I reach my highest ground

Teachers keep on teachin'
Preachers keep on preachin'
World keep on turnin'
Cause it won't be too long

Lovers keep on lovin'
Believers keep on believin'
Sleepers just stop sleepin'
Cause it won't be too long"

Sam Cooke [50]

"A Change Is Gonna Come" [50]

"I was born by the river, in a little tent
Oh, and just like the river
I've been running ever since

It's been a long
A long time coming
But I know a change gonna come
Oh, yes it will

It's been too hard living
But I'm afraid to die
'Cause I don't know what's up there
Beyond the sky

It's been a long
A long time coming
But I know a change gonna come
Oh, yes it will

I go to the movie
And I go downtown
Somebody keep telling me
Don't hang around

FOOTPRINTS

It's been a long
A long time coming
But I know, a change gonna come
Oh, yes it will

Then I go to my brother
And I say, brother, help me please
But he winds up, knockin' me
Back down on my knees

Oh, there been times that I thought
I couldn't last for long
But now I think I'm able, to carry on

It's been a long
A long time coming
But I know a change gonna come
Oh, yes it will"

It seems obvious when listening to these talented musicians sing their songs of hope, disadvantage, and protest, that their contributions to the plea for equality of opportunity is unequaled in reaching the minds and hearts of us all. Their contributions are lasting tributes and lasting expressions of the human condition and the hopes and dreams for a better world.

CHAPTER 6

Income & Taxation

Power & Privilege Prevail

Needless to say, over time taxation without representation has always won the day. So, like the 20th Century scholar H. L. Mencken, among others, is quoted as having said: "If someone says it isn't about money, it's about money."

As our financial scholars tell us, taxation is the only way in a democratic system to limit the discriminating effect of power and privilege. Progressive taxation, although under constant criticism and methods of limitation by the rich and powerful, is the only tool available to accomplish any degree of equality of income and wealth.

How can we empower our legislature and judiciary to counter unequal rights which control our government, a government which we declare to be of, by, and for the people? That question remains unanswered because our so-called government of, by, and for the people, always remains a government of, by and for those who control the electoral process with money. This control of government enables control of the distribution of money which provides the financing for a continuation of control of the electoral process, which continues to control the distribution of money - the ultimate oxymoron which feeds on itself and assures continuity of control of the distribution on money by the power of money.

This condition which prevents a truly democratic society must be addressed through our electoral system for local, state, and federal officials. And it must be addressed through our system of appointment and placement of state and federal court judges, particularly Supreme Court Justices who control our entire legal system. In essence it is revealed that tax examination, although administered by competent administrators, is far from adequate to assure fair and equitable taxation for all income levels. Some of us believe that those considered in the working class pay through taxation for the continuity of a government which sustains the dominance of the rich and powerful as they control the distribution of income and wealth.

An illustration of income and taxation rates exemplifies the essence of the distribution of income and wealth. The approximate 2019 distribution of income and taxation in the U.S. was as follows: [51]

Annual Income	% of Tax Payers	Tax Rate
Up to $15,000	9.1%	10%
$15,000 to 25,000	8.0%	12%
$25,000 to 35,000	8.3%	12%
$35,000 to 50,000	11.7%	22%
$50,000 to 75,000	16.5%	22%
$75,000 to !00,000	12.5%	24%
$100,000 to !50,000	15.5%	24%
$150,000 to 200,000	8.35	32%
$200,000 or more	10.35	32% to 37%

Allowable deductions from gross income are important in reviewing taxation as it applies to various income levels. The important determination for taxation is taxable income - total income less allowable deductions.

INCOME & TAXATION

The 2021 tax rates for various levels of income in the U.S.:

Income	Tax Rate
Less than $10,000	10%
$10,000 to 40,000	12%
$40,000 to 86,000	22%
$86,000 to $164,000	24%
$164,000 to 210,000	32%
$210,000 to 523,000	35%
$523,000 or more	37%

One could wonder what a difference it would make in achieving a more equitable distribution of income if the above chart looked more like the following chart or similar chart of a more equitable taxation schedule:

Income	Tax Rate
Less than $10,000	0%
$10,000 to 40,000	5 to 10%
$40,000 to 86,000	10 to 20%
$86,000 to $164,000	25 to 35%
$164,000 to 210,000	35 to 40%
$210,000 to 523,000	45 to 50%
$523,000 to $1,000,000	55 to 60%
$1,000,000 or more	65 to 70%

Under such a schedule the rates could apply to each level of income separately, increasing for each step-up of total income. For example: income of $10,000 or less would not be taxed, income up to $40,000 would be taxed at 10%, the next $40,000 would be taxed at 20%, the next $80,000 would be taxed at 35%, the next $50,000 would be taxed at 40%, the next $300,000 would be taxed at 50%, the next $500,000 would be taxed at 60%, and any additional income over $1,000,000 would be taxed at 70%. Overall, the total taxation would increase by 30% or more while the low wage workers would retain a greater of their earnings and thereby increase economic performance.

Economists who study and propose taxation policy could serve us well by proposing similar income tax schedules to the federal government. The additional dispensable income of the working class would have a positive effect on the economy? It is likely that most of the funds available for high income earners after taxation would not be used to stimulate the economy and create more jobs under any tax schedule. However, with more disposable income on the part of the wage earners, the economy should have an immediate beneficial stimulus as that additional expendable income is plowed directly back into the economy.

A comparison of hourly earnings at the various income levels based upon working 40 hours per week would look something like the chart below:

Income	Hourly Rate
Less than $10,000	Less than $5
$10,000 to 40,000	$5 to $20
$40,000 to 86,000	$20 to $40
$86,000 to $164,000	$40 to $80
$164,000 to 210,000	$80 to $100
$210,000 to 523,000	$100 to $250
$523,000 to $1,000,000	$250 to $500
$1,000,000 or more	$500 or more

During the aftermath of the Great Depression in the 1930s the top tax rate was said to be about 90%. Instead of this being an adverse effect on the economy it is considered by economists to have provided an effective economic stimulus in those troubled times which provided government funding without an adverse effect on the economy, which was bolstered by the spending of net income by the working-class.

Reviewing the above figures begs the question: What benefit to society makes the value of one worker $5 per hour and another worker $500 per hour? This question has been argued for many decades and perhaps centuries but lacks agreement among the classes. The answer, of course, is that those who own the gold make the rules and set the rates. An entrepreneur is simply not going to pay any more than necessary to hire and retain qualified workers. As long as the supply of workers exceeds the demand, this will not change without enforceable laws. However, the possibility of those laws coming about is very unlikely since the workers lack leverage in most industries.

The taxation policies provided by the Internal Revenue Service provide the details for a method of taxation which is intended to be fair and equitable at various income levels based upon ability to pay. Some of the provisions are detailed below.

Schedule 1: Added Earnings

Form 1040 asks you to report earnings on Schedule 1. These include: [52]

- Wages and salaries
- Business income or loss as calculated on Schedule C
- Alimony received by divorce decree or agreement entered into prior to 2019
- Taxable credits, offsets, or refunds from state and/or local tax returns
- Rents royalty income as calculated on Schedule E
- Farm income or loss as calculated on Schedule F
- Capital gains or losses
- Unemployment compensation
- "Other Income," which can include prizes and awards, gambling winnings, and earnings from an activity not engaged in for profit, such as money you made on your hobby.

Deductions from gross income allowable under current taxation rules:[52]

Adjustments to income are entered in Part II of Schedule 1. These are the amounts that were previously referred to as "above-the-line" deductions because they appeared on the first page of the tax returns that were in use in 2017 and earlier years. They were entered just above those forms' final page on the line that showed adjusted gross income.

These adjustments/deductions include:

- Educator expenses
- Costs incurred by military reservists, performing artists, and fee-based government officials
- Health savings accounts (HSAs)
- Moving expenses for members of the Armed Forces
- Several self-employment costs, such as retirement plan contributions, health insurance premiums, and half the self-employment tax reported on Schedule SE
- Savings withdrawal penalty amounts
- Student loan interest
- Tuition and fees educational expenses
- The traditional IRA deduction
- Alimony paid pursuant to decrees dated 2018 or earlier

Advantages of utilization of business income and expenses in reducing individual taxation

Depreciation of assets purchased or developed in a business operation is often declared as business expense in calculating business net income. Depreciation of expenditures which may enhance one's lifestyle, such as: automobile expenses, health care coverage, travel expenses, purchase of necessities unrelated to business operations, educational expenses, home repairs, and others, are commonly included as business expense although they may be primarily personal

INCOME & TAXATION

necessities or pleasures unrelated to business operations. The ability of the Internal Revenue Service to provide equitable limitation of such deductions to income is limited by their tax return review capability, which is limited by the budget restraints. The question of the cost to society of ignoring inappropriate reduction of taxable income is one worth exploring more thoroughly to make taxation policy more equitable.

How effective is the tax review process by the IRS?

Verification by the Internal Revenue Service of appropriate deductions is considered a primary function of tax return audits. However, the likelihood of such an audit is remote except for egregious and obvious cases of such deductions due to the enormous task of millions of tax returns being filed in a short period of time.

The number of tax returns submitted to the IRS is substantial and the ability to effectively review any large percentage of returns exceeds the capacity of the department. It has been proposed and argued that an increase of this capacity would pay dividends by discovering and/or discouraging inappropriate deductions by tax payers and those who prepare tax returns professionally.

Statistics for federal tax examiners, collectors, and revenue agents for 2020 are as follows: [51]

Number of tax examiners - 56,900
Work experience required – none
Training – Minimal
Typical education – College degree
Annual wages - $55,640
Wages per hour - $26.75

The IRS processed more than 240 million federal tax returns and supplemental documents in 2020. More than 195 million of these returns and other forms were filed electronically which represented 81 percent of all filings.

FOOTPRINTS

The 56,900 examiners to review 240 million tax returns would represent about 4,000 returns per examiner. Of course, there must be computerized selection of those triggering critical examination based upon gross income and net income. Given the limited timeframe for responsible response to taxpayers, an in-depth review of returns would represent an impossible burden for the IRS. Thus, the likelihood of a careful and complete review of any individual return is favorable for clever tax cheats.

Some returns are selected for examination on the basis of computer scoring. Computer programs give each return a numeric score. The Discriminant Function System (DIF) score rates the potential for change, based on past IRS experience with similar returns. The Unreported Income DIF (UIDIF) score rates the return for the potential of unreported income. IRS personnel screen the highest-scoring returns, selecting some for audit and identifying the items on these returns that are most likely to need review.

Considerations affecting selection for review by tax examiners: [52]

- Large Corporations – The IRS examines many large corporate returns annually.
- Information Matching – Some returns are examined because payer reports, such as Forms W-2 from employers or Form 1099 interest statements from banks, do not match the income reported on the tax return.
- Related Examinations – Returns may be selected for audit when they involve issues or transactions with other taxpayers, such as business partners or investors, whose returns were selected for examination.
- Other – Area offices may identify returns for examination in connection with local compliance projects. These projects require higher level management approval and deal with areas such as local compliance initiatives, return preparers or specific market segments.

The timeframe for auditing returns and the staffing limitations make discovery of improper reductions of taxable income and avoidance of such disclosure an inexact science. Therefore, prosecution of perpetrators of illegal tax avoidance is unlikely. It would seem to assure a taxation policy which is far less than fair and equitable, one favored by high income earners and perhaps designed and legislated by them at the peril of wage-earners.

CHAPTER 7

Leveling the Playing Field

How can taxation policies be modified to provide a more equitable distribution of income?
It seems that the only methods to effect change for a more equitable distribution of income would be through taxation or minimum wage laws. The government doesn't enact legislation requiring specific salaries or wages paid to workers other than through minimum hourly wage laws. Wages are determined primarily by supply and demand for various industries and the skills and abilities required in the various trades and functions in the workplace.

Should a steel worker be paid more than an office worker? The simple answer would be yes, however, factors other than risk and ease of the job enter into the equation. Most people who are employed as office workers would typically not seek a job as a steel worker. Office work is typically clean and safe. Steel work is typically dirty and dangerous. Office work can be done by most anyone regardless of strength or healthiness - steel work, perhaps not. Consequently, regardless of the impact of unionization, an employer can hire office workers at a lower wage than steel workers. The same comparison could be made for trash collectors, home builders, auto mechanics and many of the other similar professions.

Clerical positions such as secretarial, recording, accounting, corresponding, purchasing, sales, and other non-laborious classifications may require various skills or knowledge but are considered "white collar" jobs and are not dirty or dangerous. These positions are typically fulfilled at lower wage levels than skilled craftsman or management positions.

Management jobs in any work environment are more often filled by those with some leadership skills and decision-making abilities. So, hiring those qualified for supervisory or management positions focuses on the skills necessary to manage the functions and responsibilities for the given profession which would typically justify larger salaried income.

All of these factors enter in to the wage and salary determinations in the workplace. Hence, the wages and salaries vary considerably depending upon the qualifications of employees for various industries or governmental employers.

Taxation policy should be structured to enable the low income worker to retain a sufficient amount of earnings to support his or her lifestyle. For low wage earners perhaps the income tax should be low or none, while the high income earner, who is enjoying a much higher standard of living, should pay a much higher portion of income in taxes to support the requirements of government provided services. This, of course, won't happen by restriction of income so it must be accomplished through taxation.

The slogan "tax the rich" is difficult to dispute when considering the enormous difference in disposable income after paying for the three biggest necessities, food, shelter and clothing. The big three are a major consideration for the wage worker but of no consequence for the high income earner.

Jeffrey D. Sachs tells us in his 2012 book "The Price of Civilization," "Not a day passes without more evidence of a growing inequality of income, wealth, and power." [53]

"The vested interests still have the money and the power but have lost their legitimacy and the public trust. Big banks, big insurance companies, and big arms manufacturers are close to Congress and the White House and have successfully resisted any serious intrusions into their prerogatives."

Sachs stated: "I believe that four issues will prove to be decisive for America and its place in the world: education, environment, geopolitics, and diversity."

Any discussion or review of the effects of power and privilege should focus primarily on the only viable measure for restricting inequality of income and wealth - taxation policy. Income or restrictions thereof would be impossible to use equitably if not impossible to legislate. Taxation policy which is manipulated by the rich and powerful through deduction of various expenditures from taxable income can only be altered through legislation which is proposed, negotiated, and enacted by state and federal legislators. Of course, these legislators are elected by the generous campaign contributions to candidates for office who are cooperative with the interests of their contributors. Therein lays the difficulty of leveling the distribution of income and wealth.

Here again we are faced with the campaign finance dilemma. How much influence do the low-wage workers have on the elected officials of the local, state, and federal governments? Candidates for office expound their support for all the citizens when campaigning for election – but not for financial support. The financial support comes primarily from the rich and powerful. What they seek from the general public is votes, without which they cannot win. When they have been elected or reelected they go about the business of legislating which is where campaign finance enters the acknowledgement and repayment mode. The big contributors have a ready ear of legislators regarding matters which affect their own interests. Those low-wage workers who were sought out merely for their votes are of no further importance until reelection time, with the exception of major legislative issues which affect them directly and are well publicized.

It would be interesting to determine how legislation comes about if campaign contributions were not a factor. If the major corporations or individuals had no influence on the legislator, as is implied by campaign contributions, it would certainly change the dedication owed them for financial support. We all know that financial influence on officials is either illegal or immoral, but like they say "who's counting." Well, no one is. There is no system for tracking officials' allegiance to campaign contributors. Such a system would be impossible to administer unless someone was taping every conversation between individuals. No one knows what is said before or after elections or in private conversations. Perhaps nothing is said. Perhaps it is all just understood or implied. The elected official knows the business or interest of their supporters and if they desire reelection and need campaign funds they know to whom to pay respect and acknowledgement.

Leveling the playing field of income and taxation is such a large and multifaceted endeavor that any method of alteration for a more equitable distribution system would be impossible without complete restructuring of our system of taxation. Those with the power of the purse have manipulated our system of taxation through our system of campaign finance. If campaign contributions were unlawful, either from others or from personal funding, we would have an entirely different structure of government. If campaigns were financed on an equitable basis by government regulation and government funding and restrictions of spending were legislated and enforced we would have a different government. If government official had no means of favoring campaign contributors we would have a different government.

The facts above are hard to refute but even harder to change. How can we convince our legislators to restrict their source of funds which enables their retention of office? How can we convince those who were elected to office by virtue of campaign contributions to restrict campaign contributions? That would be like voting to ban automobiles although that is the way we get to work. It would be like asking the corporate moguls to cut their humongous salaries for the benefit of underpaid workers. That would be like asking corporate tycoons to destroy their grip on the legislative process. That would be like asking Wall Street tycoons to give up one of their vacation homes or one of their private jets for the benefit of a more equitable system of taxation. That would be like asking the Las Vegas casinos to give their patrons a better than even chance of winning. The chances of all this would be two, slim and none.

And the thing that makes all this seem unfair and ridiculous and makes having a more equitable distribution of income and taxation a menacing and troubling challenge, is that campaign finance exists as unopposed policy by our lawmakers who depend upon campaign finance to retain their positions of power and privilege.

This reminds us of the dilemma of the chicken or the egg. If you eat the chicken you won't get another egg. If you eat the egg you won't get another chicken. Which one shall we eat? Another oxymoron!

CHAPTER 8

Political Campaigns

AS WE HAVE been discussing, power and privilege are validated and legitimized by those that we elect to office who aspire to have an impact on the process of making and enforcing the rules and regulations which affect taxation. Gaining such a position is once again a function of the power of money which is established and controlled through this process.

The numerous methods of promoting candidates for elective office include written publications, broadcast media, public speeches, email campaigns, door-to-door canvassing, and personal contact by candidates – all enhanced by the determination of aggressive campaign workers and financed and made possible by political contributions primarily from the rich and powerful.

The subject of campaign finance in the United States includes the financing of electoral campaigns at the federal, state, and local levels. At the federal level, campaign finance law is enacted by Congress and enforced by the Federal Election Commission. In the states, counties, and cities it is enacted by the various legislative bodies. All of these legislative bodies are controlled by representatives who are elected with the support of campaign contributions on a large part by those who can benefit by faithful acknowledgement of their generosity.

Campaign expenditures have grown enormously over the last few decades, particularly since corporations have been granted the right to make unlimited contributions to political action committees (PACS) in political campaigns. The decision by the Supreme Court in the Citizens United case was a victory for corporate interests and to the demise for those who declare that corporations are not people and that granting them the right to finance PACs grants corporate officers the right to place their thumb on the scales of equal rights of all the people.

Print Media – Newspapers

Written publications consist primarily of local, state, and national news publications such as the USA Today, Wall Street Journal, New York Times, Washington Post, Boston Globe, Chicago Tribune, Los Angeles Times, and many other publications of major cities which are distributed in surrounding areas. Advertising rates vary depending upon the total distribution of various publications and the demand for advertising space.

The cost of effective advertising in print media is based primarily on the distribution. The cost of advertising in the New York Times with a circulation of over a five million subscribers could amount to thousands of dollars or more depending upon the size and frequency of each placement. The Chicago Tribune with circulation of five to eight hundred thousand, and Los Angeles Times with circulation of six to nine hundred thousand would amount to lower rates. Smaller distribution areas such as Kansas City, St. Louis, Dallas, Miami, Denver, Phoenix, and Seattle would amount to yet lower rates.

The Los Angeles Times is the largest daily newspaper in the Los Angeles, CA area. The newspaper has a daily circulation of approximately 815,000. It is owned by Tribune Company. The estimated ad rate for the newspaper is $865.00, which is typically for a column inch of black and white advertising space. Seasonal factors also apply.

Getting the raw product, editing it to your preference, and implementing it in the final ad has an average cost of $113,000 for major newspapers. Then you must add the cost of placement.

Print Newspaper Approximate Advertising Cost [54]

Publication	Circulation	Full Page Ad
New York Times	5,500,000	$150,000
Los Angeles Times	650,000	$70,000
Washington Post	550,000	$160,000
Chicago Tribune	500,000	$100,000
Kansas City Star	250,000	$150,000
Milwaukee Journal	150,000	$25,000
Bozeman Daily	15,000	$3,000

These price comparisons are estimates for illustration only

The majority of newspaper ads will cost much less than the prices above because newspaper ads are typically a portion of a page, rather than a full page.

In addition to print ads, most papers also offer online ads that display on their website. In a digital age, more and more readers are turning to the web to access news instead of turning pages of a physical paper. Without printing costs, digital ads are typically much less costly than print ads. Pricing can range from as little as $50 for a local paper with a small circulation to thousands for a major publication, though some only offer print or print and digital advertising options.

An effective campaign utilizing print or digital media would require frequent utilization of publications in the specific area of the political office and could consume millions of dollars of advertising costs in order to achieve promising results.

Broadcast Media – Radio and Television

Broadcast campaigns would require utilization of numerous sources in any market area as there are numerous stations which share the broadcast market on radio as well as television. Reaching a wide audience would require utilizing all of the most popular stations or channels, each one with rate structures which could amount to thousands of dollars depending upon the statistics of listeners or viewers at various times of day or evening.

In the most recent national political campaign a gargantuan sum of political ad money was expended - $8.5 billion in total. In the fall presidential contest alone, nearly $1.8 billion was spent.

President-elect Joe Biden spent $661 million, more than $1 billion inclusive of Democratic groups, compared to President Donald Trump's $500 million spend, more than $760 million inclusive of Republican groups.

Remarkably, California—a state not in play for the presidential election—was the biggest recipient of ad dollars overall. Ballot propositions and competitive House races fueled nearly $475 million of ad spending. [55]

The 2020 election cycle is by far the most expensive campaign year in history. Advertising spending on candidates running for federal office across the country totaled at least $2.5 billion.[55]

Close to 5 million ads for president, senate and congress have aired on television airwaves alone. That's more than twice the volume of ads in the past two presidential cycles.

Mailing Campaigns

Effectiveness of mailing campaigns would require repetitive mailing of the entire area of the candidate's representation. Mailing lists of registered voters could be used to limit the number in each district but would lack assurance that all potential voters would be reached.

Mail pieces can cost from about 5 cents each to more than 2 dollars each depending upon the size, type, and quantity purchased. The cost of design by political consulting firms could amount to $5,000 or more. [56]

Digital banner ads distribution would typically cost from a thousand to several thousand dollars depending upon the number of impressions.[57]

Telephone Canvassing

Outside door knocking and phone canvassing is the most effective way to get out the vote. In the most recent election, when the pandemic restricted social contact, phone canvassing became more critical.

Telephone canvassing is an expensive method of reaching all potential voters. The operation of a "boiler room" of telephone canvassers requires significant supervision as well as numerous callers who must be paid minimum legal hourly rates.

If one employed a caller at the minimum wage of $15.00 per hour the daily cost for an eight hour day before payroll taxes and benefits would amount to $120.00. If a caller can reach 10 potential voters in a positive way per hour they would reach 80 voters per day. This calculates to a cost of approximately $1.50 per contact. If these 80 contacts amounted to positive persuasion of 50% it would calculate to $3.00 each. Based on the need for 100,000 votes to win an election the cost of winning totally by phone contact would be $300,000. This, of course, would not be the case, but it provides a cost comparison with other campaign tactics.[58]

A cost effectiveness comparison of various methods of political campaign professionals indicated that phone canvassing has been the second-best way for voter outreach after door-to-door canvassing.[58]

The pricey presidential showdown between Joe Biden and Donald Trump in the 2020 election was funded by an unprecedented number of small donors giving online and billionaires who wielded tremendous political influence over the last decade. Donors also fueled record spending in congressional races.[59]

Total federal spending in the 2020 election reached $14.4 billion, the most expensive election in U.S. history by a large margin. Biden's campaign became the first to raise over $1 billion from donors. Biden's cash advantage over Trump helped him pepper swing states with far more campaign ads. Biden also received more help from super PACs and "dark money" groups. [59]

Trump's campaign raised $774 million. Trump raised over half of

FOOTPRINTS

his money from small donors giving $200 or less, a stunning figure no other presidential candidate has matched. Trump continued raising money long after news outlets called the race for Biden, racking up campaign cash he could use to influence the future of the GOP.

While the presidential election drew a record $5.7 billion, congressional races saw a stunning $8.7 billion in total spending.

Nine of the 10 most expensive Senate races and five of the 10 most expensive House races ever occurred happened in 2020. The rise in congressional spending started in the 2018 cycle, which smashed midterm spending records.

Total pending in presidential election years was less than $6 billion in 2008, more than $7 billion in 2012, more than $7 billion in 2016, and more than $14 billion in 2020, of which $5.7 billion was for presidential campaigns and 8.7 billion in congressional elections.[59]

Robo-Calling

Robo-Calling, a term used to identify automatic calling with a computer program dialing the phone numbers and upon an answer playing a recording promoting the candidate, is commonly used. The effectiveness would not be equivalent with personal calls but the cost per contact would be less. The use of such an annoying telephone campaign is banned in some areas and many believe that it should be considered an invasion of privacy in all areas.[60]

Door-to-Door Personal Contact

And, of course, personal contact of campaign workers going door-to-door to reach potential voters and encouraging votes to support a given candidate would be the most expensive method of all unless the campaign workers are volunteers who agree to provide their service without charge. Even without paying the volunteers, the staff required to canvass a given area effectively would be considerable and difficult to maintain.

The years since 2000 have seen a widespread revival of election canvassing. An intensive effort by the Al Gore campaign of 2000 was credited with gaining several points on election, enough to win the popular vote despite being down several points in the polls the day before. Subsequently, the Republicans launched their 72 Hour Program of get out the vote efforts over the last three days of a campaign, and also found demonstrable proof that it gained them several points in key races.[61]

Who finances campaigns for elective office?

Campaign finance seems unlimited, particularly since the adoption of the Citizens United Decision by the U.S. Supreme Court which gives corporations the same rights as individual citizens to finance campaigns.

It is declared in our constitution that our government is established of, by, and for the people. With the Supreme Court Citizens United decision our government is now established of, by, and for the people and the corporations. Of course, the obvious problem with such a system is that campaign finance enables a huge thumb on the scale of equality of opportunity in this country. By allowing corporations to finance campaigns, corporate owners and officers have additional votes, one as individuals and one as corporate owners or officers, not directly, but through campaign finance, which is not regulated to provide a level playing field in the influence of voters on the quality, ability, and integrity of candidates for office.

What constitutes a successful democratic system of government? Is it meant to provide equality in standards of living or accumulated wealth? Is it defined by equality in political power? Is it meant to provide a spirit that drives competitive production and distribution of goods and services? Is it a means of preventing one individual or group of individuals from garnering an unfair advantage over others?

In legislating and enforcing such a system of fairness, does it prevent dishonesty and unfair advantage? Does it prevent upper-class perpetuation of advantage at the expense of those who lack the ability or opportunity to pursue equality of opportunity? Or does it support those who attain political leadership and inordinate power over others? Does it deter unfair competition in the business environment?

All of these questions can be answered positively or negatively depending upon one's station in life and opinion about the ease or struggle for equality of opportunity, but one fact seems to be confirmed throughout a discussion of our electoral process – money talks and money elects, (or as others say – money talks and b. s. walks.) Any thoughts otherwise are merely rhetorical pipe dreams which sound like electoral equality but are just wishful thinking with no evidence for confirmation.

CHAPTER 9

Measures for Political Equality

Suggested remedies in expanding equality of opportunity:
Considering the failure of quasi-democratic forms of government in leveling the playing field of economics, measures could be taken to strengthen the ability of the working class to garner a stronger position. Some suggested methods of change include the elimination of political contributions – requiring all political campaigns to be financed through government regulation and control. Some suggest the reversal of the Citizens United decision by the Supreme Court which basically declared that corporations are citizens when making unlimited contributions to campaigns. The likelihood of such a reversal is remote at the time of this writing due to the conservative majority on the Supreme Court.

Some have suggested that we elect Supreme Court Justices by popular vote or some method through which the judicial system can select justices so that one individual, such as the President of the United States, cannot appoint judges by personal preference.

Some suggest that there should be term limits for all elective and appointive offices including judgeships in all courts including the Supreme Court, such that no elective official can be considered a career politician or elected official.

The gap in income and wealth between the wealthy and everyone else has continually widened year after year. Some of us believe that it will continue to do so until the gap between wealth and poverty causes the economy irreparable harm. Perhaps this will not change until political campaign finance is eliminated. The wealthy individuals and corporations finance political campaigns for those who enact and enforce laws for their advantage.

The enactment and enforcement of laws, rules, and regulations, particularly those pertaining to taxation, are completely controlled by the rich and powerful and that will not change until we accept it as fact and unite to change it. Of course the obstacle to this coming about is inevitably the lack of time and effort by the workers to initiate change through the legislative process while working fulltime jobs, and often more than one job, to provide for livelihood and education of their families. This condition is exacerbated by the inequity of compensation between workers, management, and ownership.

What methods of promoting and evaluating the breadth of knowledge, personal traits, and political policy attributes of candidates for elective office could replace the power of money in promoting political candidacy - both personal campaign spending and organized campaign funding "

Like Robert Scheer told us in his 2010 book: "The Great American Stickup": "The influence of big corporate money overwhelms that of labor, environmental, consumer or grassroots organizations, making a mockery of the American ideal of self-government when it comes to reining in the antics of the largest conglomerate of wealth." [62]

These are the challenges to providing more equality in our electoral system. It continues to be all about money. It simply cannot be all about money and be a fair and equal system of government. Will that change? Not until campaign finance is controlled or eliminated, an unlikely solution which will continue to be challenged by our "free speech-focused" Supreme Court justices. A majority of these justices were appointed by conservative leaning presidents. That plurality is unlikely to change with campaign finance laws as they now exist. But, the imperativeness of such change to assure some measure

MEASURES FOR POLITICAL EQUALITY

of equality of opportunity in future elections reinforces its urgency.

Jeffrey D. Sachs tells us in his 2012 book "The Price of Civilization," "... for thirty years, tax increases have been vilified and rejected at the polls. If that continues, America's days as a global leader and prosperous economy are numbered. Initiatives to upgrade the infrastructure and improve education for the poor have been crippled by inadequate budgets." [53]

Sachs, in his comments, senses a readiness for support to raise the tax burden on high income households. Considering this factor, along with the governing majority trending younger and more progressive, especially with the increase in African Americans and Hispanics in the voting population, we should note that his comments were cited five or more years ago, and we have as yet seen no indication of any significant change in attitude toward taxation policies.

Campaign finance is nothing more than a tool for the extension of power and privilege. It enables those who use power and privilege to enhance their enjoyment in life, their dominance over those less fortunate, those less focused on the advantages of income and power, those denied a birthright to the wealthy class. It enables those in the advantageous position of wealth to manipulate the laws and the enforcement of laws to their advantage to protect and enhance their wealth and thereby their power and privilege.

How do they retain such a system we ask? The answer is simple – with money. Who gets invited to the limited political events? Who gets their phone calls to public officials and office holders returned? Who shakes hands with the public officials? Who gets a nod and a wink from a public official? And the answer is, those with the power of money. Has it happened to you lately? Has it happened to you ever? Not for most of us. Not for most of those who are interested in reading this book.

There have been pleas for campaign finance reform for decades, particularly the last few decades. The pleas come from both parties, particularly the party out of favor at the time. And what do these pleas engender? You guessed it - nothing. The financiers of the political campaigns simply cannot allow this to happen. Without campaign finance they have no leverage on those seeking or trying to retain office. It is like what Willie Sutton was quoted to have said many years ago when asked why he robs banks: "Because that's where the money is."

The best time to change the electoral system would have been 1787. The next best time would have been in every session of the court since 1787. The next best time is 2022. The only way to change from a system where we have "the best government that money can buy" is to amend the constitution. This, of course, requires a constitutional amendment, which requires proposal by Congress through a joint resolution passed by a two-thirds vote, or by a convention called by Congress in response to applications from two-thirds of the state legislatures. Eleven thousand amendments have been proposed since 1989 and twenty-seven have been enacted.

An amendment proposed must be ratified by three-fourth of the states. The last time a propose amendment gained the necessary two-thirds votes of the House of Representatives or the Senate for submission was in 1978. It expired without the ratification by the states, as required by law.

So, what are the chances of a two-thirds vote in Congress or the Senate? Neither party has the polarity in either house to achieve the vote required. What are the chances of applications from two-thirds of the states? Neither party has a plurality of support in enough states to gain the requirement. Neither of these actions would be impossible but neither would have a chance better than "slim or none" of coming about in the legislative bodies we are strapped with today.

CHAPTER 10

Footprints on a Struggling Planet

As we all know by now, the Earth has been gradually heating up since heavy industry began to blacken our skies not long after civilization came about. Now that trend is accelerating as larger chunks of polar permafrost dissolves in the sea and the sea level continues to rise and creates floodwaters in many parts of the planet. Such danger has been ignored or denied by corporate propaganda for decades but now it has been confirmed and there is urgency to place blame, point fingers, and toss around solutions which will fail to make a difference and are unlikely to happen.

What is our footprint on the planet? What is our evidence of having been here? Our footprint as a people would begin with the inception of the human life. It would include the rapid expansion of humanity and the invention and the development of goods, services, and ways of life we have created. It includes our obsession, consumption, and waste of vital resources the Earth provides, some of which may replenish, but some of which will not. It would represent the heating of the planet due to fossil fuel consumption which exhausts carbon dioxide into the atmosphere and forms a layer of gases around the planet which allows heat from the sun to enter but restricts it from escaping from our atmosphere, remaining here to endanger the life-sustaining necessities for all life forms.

The threat of such painful outcomes induces many questions. What has been the impact on the survival of humanity by the footprint of all mankind? How much has our footprint increased the impact on the longevity of human necessities? Are the necessities of human life being depleted beyond restoration for future generations of humans? Will scientific discoveries and developments be capable of providing a continuity of human life for centuries to come?

The incessant increase of world population provides reason for concern for our environmental scientists. Although current projections indicate a future decline in population increase and a possible decrease, this also has a negative factor to consider. The ability of a declining population of working-age people to provide sustainable necessities of life may be the most important and imminent consideration. Could a decline of the birthrate create an unexpected dilemma? If we must reduce world population to save the planet's ability to support human life, and by doing so are unable to provide the necessities of life for the smaller population, we may be facing an outcome such as the cliché of the medical world: "the operation was a success but the patient died."

Many of our scientists are no doubt burdened with this question in searching for an outcome which would assure a continuity of mankind. To date there seems to be no general consensus of opinion on the subject. One primary consensus focuses on the fact that the prevention of an unfavorable outcome will require significant changes in lifestyles on the world population. This may be a hard sell in this country, where we are obsessively focused on comfort and pleasure at all times, and around the world as those less-developed countries continue to strive for lifestyles comparable to ours, while simultaneously we continue to expand our lifestyles beyond the capability of the planet to replenish what we consciously and selfishly consume or destroy.

The chances of curtailing such overconsumption and wastefulness of precious resources do not look promising. The dissemination of information and encouragement to restrain unnecessary use is almost nonexistent. Information and warnings which are provided seem to take little notice and receive little distribution. It is like deciding not to tell a child about the big bad wolf behind the tree because it might create fear even though it may save the child's life. In this case the fear might have an adverse effect on the vibrant economy, a condition which the rich and powerful and those of us who strive to be such prefer to avoid for the sake of maintaining the order of things to maintain their grip on the power of money.

Speaking of this seemingly impervious struggle for equality of opportunity, power, and privilege - a simple statement attributed to Ralph Waldo Emerson justifies the challenge: "The voyage of the best ship is a zigzag line of a hundred tacks." In other words, you can't just point it in the right direction and expect it to reach the destination – there are too many events along the way which can deter the best of all travel plans – ocean waves, massive storms, high winds, and human error.

Thus, the struggle for equality of opportunity faces obstacles other than having great ideas and working hard. It also may require having the proper parents, achieving a position of power and influence, or being in the right place at the right time.

Working hard and being a dedicated employee does not assure one of success on the ladder of employment hierarchy. Politics and timing, which are not available as equality of opportunity measures, are often as or more important in advancement and are not provided by a level playing field. It is obviously easier to run downhill than uphill, and the uphill battle for success is a steep climb without political power and good fortune.

It seems that morality is a trait without support in this game of chance. C. Wright Mills discussed in his book "The Higher Immorality" of the 20th Century. He suggests that immorality is not merely a matter of corrupt men but is a systemic feature of the American elite, a general acceptance of an essential feature of the mass society. He tells us that there are not only corrupt men but the institutions are corrupting them. Mills states that it is not merely a question of a corrupt corporate or state administration, but is a feature of the corporate-rich deeply intertwined with the politics of a military state. This is not new information. No, Mills warned of this more than 50 years ago.[63]

As we digest the revelations of C. Wright Mills we may recall the many questions of integrity on the part of government officials which are publically disclosed, discussed, and criticized and then disappear from public view as fast as they appeared – forgotten and dismissed as part of the process and considered not of significant importance in the process of governing. How often do these charges sort of end up with explanations and resolutions such as a child caught with his or her hand in the cookie jar and saying: "I didn't do it - and - I'm not going to do it again."

My wise old friend Skip Sleyster, may he rest in peace, called these types of officials "cookie fingers." Skip feared not and called politicians like he saw them. He could do that because he and his friend Leo Hallak made money doing what others avoided, living a meager lifestyle and buying industrial real estate sites at bargain prices that others considered unattractive and recycling them for productive use by upcoming entrepreneurs.

And as the comedian, George Carlin, used to tell us when talking about politicians' comments when charged with some malfeasance in office: "I just want to get this thing behind me and get on with my life," or: "This is taken out of context," or: "It has been suggested that...," or: "This mere oversight on my part ...," or: "This attempt to question my integrity ..., or: "We want to move forward to meet these challenges, or: "I'm taking responsibility for my actions," or: "This is merely a miscommunication," or: " I will respond to (not answer) these charges," or the last resort: "Whatever happened to innocent until proven guilty?"

It seems that the incessant march of the "petroleum at any cost" segment of our society continues unabated as the professional polluters continue to pump oil from its burial grounds of centuries ago and as it continues to destroy humanity one gallon at a time as we ignore the greatest threat to humanity that exists today.

One fact is certain: Government of, by, and for the people, all the people, is neither practical nor possible without the elimination of the power of money on the scales of equality in the electoral process. Such a statement of a candidate being "the best that money can buy" must be eliminated as a slogan, whether meant as praise or slander. It is usually meant as slander.

Of course, eliminating the power of money will be challenged by our other rights and privileges, such as the right of free speech. This was exemplified by the recent Supreme Court decision in the Citizens United case, in which it was declared that corporations can engage in electoral finance as if they were people. In other words, they have the right of free speech and spending unlimited amounts to elect your favorite candidate is free speech. In the Supreme Court decision that gave the corporations this right to sway our elections with their power of money, Justice John Paul Stevens, in his dissenting opinion stated that "the conservative majority had ignored common sense as well as a hundred years of settled law." This simply and sadly gives those who own and control a corporation additional voting power in electing and maintaining those favorable to their interest in positions of power. How democratic is that? What are the options of those unable to finance campaigns to have equal rights in the electoral process? Obviously, there are none. "If they say it's not about money, it's about money."

Mark Crispin Miller asks us in his Foreword for the book "Screwed" by Thom Hartman: "How did America become the place it is today – a quasi-gulag of bright shopping malls and hidden torture chambers of crumbling schools and sprawling private jails? [64]

Well, some suggest that we continue to slide down the path of unbridled capitalism for the few and the grips of unrelenting oligarchy for the rest of us.

Thom Hartman tells us that so-called Republicans have established a mind-numbing record of polluting the environment; bloating government; appointing crony partisans; pushing the nation into debt to fund tax cuts for the rich; legislating catering to the world's largest corporations; opposing women's rights; kneecapping states, local communities, and schools; evisceration of constitutional protections of liberty at home; and devastating our nation's reputation abroad." [64]

How can an individual or a group of individuals compete with the big corporations when it comes to financing electoral campaigns? This is another rhetorical question. It answers itself. They can't. Corporations, while assuring themselves substantial profit, include every outlay of capital as an expense of some type. They price their goods and services adequately to assure a healthy bottom line, otherwise the executives would soon be replaced from their lucrative positions, which are loaded with huge salaries and perks, so that others who are more receptive to legal bribery of politicians would be available to get the taxation and regulatory provisions which help to assure continuity of their grip on the populous.

Some would suggest the "free press" as a solution. This can be discounted without discussion because the free press is not free. All of the major publishing and news organizations are owned and controlled by rich and powerful individuals and corporations. Submitting articles to be published therein is a waste of time. They all have their own experts, writers, and prognosticators who follow the political slant of the publication. What are the chances of your article or my article becoming published if it might be adverse to the beliefs of the publisher? This would once again fall under the dreaded cliché of "slim and none."

It appears that attempts to remove the thumbs of the rich and powerful from the scales of justice is a fruitless endeavor under our undemocratic electoral system. Money is allowed to control the electoral process. And it appears that such a system is gaining in strength and survivability by virtue of the cooperation of the federal courts in supporting the desires of those who appointed them. Appointments of federal court judges are made by political leaders selected with the power of money. As our Supreme Court justices have made quite clear, corporations are the same as people when it comes to political campaign finance, a condition which assures a government of, by, and for those with the power of money – not as our founding documents state: "of, by, and for the governed."

The fact that we often are either unaware of or fail to acknowledge is that holding political office is a profession in and of itself. Those who hold political office are not primarily in it due to an urge to serve humanity for a higher purpose. That may have been part of their original intent for seeking office but they aren't totally consumed by a desire and drive to come to the rescue of the poor and needy. They have no ongoing intention or drive to level the playing field of income and wealth.

No, holding public office is a profession loaded with a living salary and lots of perks. These perks include: air travel privileges and priorities, premium health insurance, access to the news media, public awareness, special privileges in one's home state or city, access to the best seats in the house, and open doors anywhere one wishes to go. Some office-holders are blessed with unspoken financial opportunities which often come with spoken or unspoken obligations to support political goals and special privileges of such major campaign contributors. Politicians who fail to realize and acknowledge this practice sometimes have very short political careers.

We, the voters, are fed a long list of achievements and qualifications of perspective candidates for office. Most voters have little or no time to confirm such claims or to assess the proficiency of such candidate to perform the duties of the office which they pursue. The claim of superior integrity, which is an assumed standard quality for election, is either disregarded as a basic character trait or is assumed without verification. The common character traits of a candidate are, of course, proclaimed managerial expertise and accomplishments, and concern for the common good and equality of opportunity.

The lack of financial backers and established references is a major obstacle for unconnected potential candidates to seek office even if he or she is highly qualified and of highest integrity.

FOOTPRINTS

It seems that we can repeat with more assurance what H. L. Mencken, among others, is quoted as having said about political matters: "If they say it's not about money, it's about money." Such a belief is evident as we review the business world, which is controlled and manipulated by those who own the gold, those with power and privilege – the rich and powerful individuals and corporations.

To level the playing field we have to change the rules. To change the rules we have to have a government of, by, and for the people - all the people, not just the rich and powerful. To have that we have to get money out of politics, something we lack the power to do without an uprising of the populace. But such an uprising is unlikely without the populace gaining control of the media. And, this is unlikely to happen because the media is now owned and controlled by those with power and privilege. What we have here an oxymoron with no apparent solution.

The purpose or intent of this writing is not to question or oppose the free enterprise system. There is no intent to suggest a system whereby everyone is granted the same standard of living or enjoyment of life. There is no intention to suggest that everyone receive the same remuneration from their profession. Our financial system should be free to accomplish these goals, but with regulations to enable a level playing field.

The purpose of this writing is to question the system of campaign finance which has an adverse effect on equality of opportunity and equal rights under the law. They say that money won't buy happiness but it sure will take some pressure off, particularly when used to influence our elected officials.

The size of our footprint in the sand shouldn't be enhanced because of our station in life or those we may have bribed along the way, particularly elected officials who pay for financial support with perks and privileges. It should be enhanced or limited by our honesty and integrity with those we encounter, including those less fortunate. It seems that helping those less fortunate must require steps and policies to enable or enhance such an achievement. And it seems that this can only be accomplished by taking the influence of the power of money out of the equation. And it seems that this will require negating the erroneous decision of the Supreme Court Justices, those who are revered by the leading Republican legislators and who believe that corporations are people and that money can vote.

Our political system seems like a poker game. You have to have the ante to enter the game, the one with the most chips wins, there aren't any re-deals, and when your chips are gone you must leave the table. The only thing that matters is money. It is distributed during the game based upon the skills and cleverness of the players – and upon luck of the draw. Is it fair? The answer to that varies by the outcome of the players. Is it fair for one player to bring more money to the table if his or her holdings disappear? Is it fair for one player to borrow from another player? All these rules and/or practices of our political system are established by the players. Without substantial funding one cannot play in the game of politics. Therefore, most of us are unable to play the game.

CHAPTER **11**

Pros and Cons of Equality

ECONOMIC INEQUALITY HAS been an on-again off-again topic amongst those in the financial services world. The Occupy Wall Street movement was one notable effort to shine light on a subject which is rendered unimportant by many of the financial tycoons of the corporate world.

Occupy Wall Street was a protest movement against economic inequality that came about in 2011 in the New York City Wall Street financial district. The issues were economic and social inequality, greed, corruption, and undue influence of corporations on government, particularly in financial services. [65]

The protesters originally occupied Zuccotti Park. When forced out the focus reverted to occupying banks, corporate headquarters, board meetings, foreclosed homes, and college and university campuses.

The slogan "we are the 99%" referred to the wealth inequality between the wealthiest 1% and the rest of the population.

During the occupation in Liberty Square, a declaration was issued with a list of grievances. The declaration stated that the "grievances are not all-inclusive.

Occupy Wall Street's goals included a reduction in the influence of corporations on politics, more balanced distribution of income, more and better jobs, bank reform (especially to curtail speculative trading by banks), forgiveness of student loan debt or other relief for

PROS AND CONS OF EQUALITY

indebted students, and alleviation of the foreclosure situation. Some media labeled the protests "anti-capitalist", while others disputed the relevance of this label. [65]

During an October 6 news conference, President Barack Obama said, "I think it expresses the frustrations the American people feel, that we had the biggest financial crisis since the Great Depression, huge collateral damage all throughout the country ... and yet you're still seeing some of the same folks who acted irresponsibly trying to fight efforts to crack down on the abusive practices that got us into this in the first place." How this squares with the immediate bailing out of the banks is a question debatable infinitum.

The movement goals included reducing the influence of corporations on politics, more balanced distribution of income, better jobs, bank reform, and forgiveness or relief of student loan debt.

The Occupy Wall Street campaign announced that it had wiped out almost $4 million in student loans, amounting to the indebtedness of 2,761 students. [65]

Occupy Wall Street has been credited with reintroducing a strong emphasis on income inequality into broad political discourse and, relatedly, for inspiring the fight for a $15 minimum wage.

Pros and Cons regarding raising the minimum wage

In reviewing the issue of the minimum wage the pros and cons are what one would expect from the standpoints of the workers and the financial gurus of Wall Street.

The workers' interest is obviously a living wage, one which will keep pace with inflation and cover the ever increasing cost of housing, food, healthcare, and other goods and services, most of which is owned and provided by investors.

The financiers' interest is of course return on investment, appreciation in real estate values, and profit margins on the goods and service provided to the populous.

The federal minimum wage was introduced in 1938 during the Great Depression under President Roosevelt. It was initially set at

$0.25 per hour and has been increased by Congress 22 times, most recently in 2009 when it went from $6.55 to $7.25 an hour. Many states plus the District of Columbia have a minimum wage higher than the federal minimum wage. [66]

Proponents of a higher minimum wage state that the current federal minimum wage of $7.25 per hour is too low for anyone to live on; that a higher minimum wage will help create jobs and grow the economy; that the declining value of the minimum wage is one of the primary causes of wage inequality between low- and middle-income workers; and that a majority of Americans, including a slim majority of self-described conservatives, support increasing the minimum wage. [66]

At $7.25 per hour a worker would gross less than $300.00 for a week's work. After Social Security taxes and other mandatory deductions that would probabl1y amount to $250.00 or less. Basic expenses of food and other necessities would then reduce that amount to less than $200.00 to cover the cost of clothing, transportation, housing, insurance, medical care, and other necessities. Where can one obtain or maintain housing for less than $200.00 per week? In California and New York, that would mean living in a tent in some unauthorized area without heating or sanitary facilities. It is impossible without working more than one job or having more than one income per household.

Because the federal minimum wage is not indexed for inflation, its purchasing power (the number of goods that can be bought with a unit of currency) has dropped considerably since its peak in 1968. The minimum wage in 1968 was $1.60, which is equivalent to $11.16 in 2016 dollars and which is 53.9% higher than today's $7.25 federal minimum wage. Between July 2015 and the last increase in the minimum wage in 2009, the federal minimum wage lost 8.1% of its purchasing power to inflation. According to Liana Fox, PhD, Senior Analyst at the Economic Policy Institute, "inflation indexing guarantees low-wage workers a wage that keeps pace with the rising costs of goods and services." Raising the minimum wage and indexing it to inflation would ensure that low-wage workers could adopt a standard of living commensurate with the current economy. [66]

Opponents say that many businesses cannot afford to pay their workers more, and will be forced to close, lay off workers, or reduce hiring; that increases have been shown to make it more difficult for low-skilled workers with little or no work experience to find jobs or become upwardly mobile due to decreased hiring. They claim that raising the minimum wage at the federal level does not take into account regional cost-of-living variations where raising the minimum wage could hurt low-income communities in particular.

The Economic Policy Institute states that a minimum wage increase from the current rate of $7.25 an hour to $10.10 would inject $22.1 billion net into the economy and create about 85,000 new jobs over a three-year phase-in period. Economists from the Federal Reserve Bank of Chicago predicted that a $1.75 rise in the federal minimum wage would increase aggregate household spending by $48 billion the following year, thus boosting GDP and leading to job growth. [66]

The Congressional Budget Office projected that a minimum wage increase from $7.25 to $10.10 would result in a loss of 500,000 jobs. In a survey of 1,213 businesses and human resources professionals, 38% of employers who currently pay minimum wage said they would lay off some employees if the minimum wage was raised to $10.10. 54% said they would decrease hiring levels. San Francisco's Office of Economic Analysis said that an increase to $15 would reduce the city's employment by about "15,270 private sector jobs." [66]

60% of small-business owners say that raising the minimum wage will "hurt most small-business owners," according to a 2013 Gallup poll. Jamie Richardson, Vice President of fast food chain White Castle, said that the company would be forced to close almost half its stores and let go thousands of workers if the federal minimum wage were raised to $15. Forbes reported that an increase in the minimum wage has led to the closure of several Wal-Mart stores and the cancellation of promised stores yet to open. [66]

A study from the Federal Reserve Bank of Cleveland found that although low-income workers see wage increases when the minimum wage is raised, "their hours and employment decline, and the combined effect of these changes is a decline in earned income... minimum wages increase the proportion of families that are poor or near-poor." [66]

Let's face it. The federal minimum wage should be increased substantially and immediately. In Los Angeles the minimum wage is $15.00 per hour which still makes it difficult-to-impossible to cover housing after the other life necessities are funded.

The argument of business owners is that higher wages mean higher costs of production. Higher costs of production mean higher prices. Higher prices mean smaller quantities of goods and services sold, which means less demand for the number of workers needed for production. This is what we call an oxymoron.

Jeffrey Sachs addressed this subject matter in 2012 in "The Price of Civilization." Paraphrasing his comments: "At the root of America's economic crises lies a moral crisis: the decline of civic virtue among America's political and economic elite, markets, laws and elections are not enough if the rich and powerful fail to behave with respect, honesty, and compassion toward the rest of society. Without an ethos of social responsibility there can be no meaningful and sustained economic recovery." [53]

Sachs sums up his message: "We have great tasks ahead, to redeem once again the American trust in democracy and equality. We have a high responsibility to our children and other generations that will come."

Sachs seems to point out the fragility of sustainability of our system of free-enterprise which lingers due to egocentric attempts at garnering all the wealth into the hands of the few. He seems to focus on equality of opportunity as the essential ingredient in any democratic or quasi-democratic system of government. In his words: "Our future lies in a healthy, productive balance of competition and cooperation in an interconnected society." [53]

Bill Moyers – "Moyers on Democracy"[67]

Bill Moyers, organizer of the Peace Corps and White House assistant to Lyndon Johnson in the 1960s, discussed this subject in 2009 in his book "Moyers on Democracy." [67] He tells us that we have fallen under the spell of money, faction, and fear in pursuit of wealth, power, and empire. He tells us that the earth we share as our common gift, to be passed on to our children's children, is being despoiled as private wealth is growing and public needs increase and that the conclusion that we are in trouble is unavoidable. Moyers mentions coal mining tearing the tops off mountains and dumping them in the rivers and the stock-market scorning long-term investment in favor of short-term turn overs leaving insiders with stuffed pockets as millions of small stockholders and pensioners out of luck and hope. And he discusses the firewall regulations that were removed for the benefit of speculators who reward the legislators with generous campaign contributions. [67]

Moyers quotes Mary Elizabeth Lease more than 100 years after the Constitution was signed: "Money rules - Our laws are the output of a system which clothes rascals in robes and honesty in rags." Her words are as true today is they were then. She tells us that when the state becomes the guardian of power and privilege to the neglect of justice for the people as a whole, it mocks the very concept of government as proclaimed in the preamble to the Constitution and mocks the democratic notion of government as a voluntary union for the common good.

Jacob S. Hacker and Paul Pierson - "Winner-Take-All Politics"[68]

Jacob S. Hacker and Paul Pierson make the point loud and clear in their revealing 2010 book, "Winner-Take-All Politics, How Washington Made the Rich Richer and Turned Its Back on the Middle Class." They point out the shocking and oft-ignored abusive distribution of income between the executives and the workers who make it all happen. They tell us that "in 2009 the investors and executives of the 38 biggest companies on Wall Street earned $140 billion and that Goldman Sachs paid its employees nearly half a million dollars each." They tell us that "the CEO of Goldman Sachs was paid $68 million in 2007 and the top 25 hedge managers made $892 million dollars on average." To make this even more disturbing, "it occurred during the time when ordinary workers struggled in the worst downturn since the Great Depression. And to make matters worse again, Wall Street was thriving because two years before they received hundreds of billions of dollars in federal bailout money." And "they were bailed out from a crisis that they created by their own reckless practices." [68]

Hacker and Peirson tell us that "… those who have the most power in the market also have the most power in politics, undermining the basic ideal of which democracy rests. All the greatest theorists from Aristotle to Alexis de Toucqueville, from Plato to Thomas Paine – have expressed concern about the sizable gaps in economic standing. Alexis de Toucqueville, who lived from 1805 to 1859, is quoted as having stated that "a government of a democracy is the only one under which the power that votes the taxes escapes the payment of them. In other words, the less affluent majority could use its political power to expropriate the rich. He made this statement about 200 years ago. [68]

They remind us of an alarming revelation of "the 2000 platform of the Texas GOP, the state party of George W. Bush, Tom Delay, Dick Armey, and Phil Gramm. It called for a return of the gold standard, the abolition of the Federal Reserve, the elimination of the minimum wage, the abolition of Social Security and the repeal of the Sixteenth Amendment, which created the federal income tax, and the elimination of the IRS." [68]

This, of course, begs the question: How can so many so-called intelligent and well-meaning people have such deranged beliefs about democratic programs which make our economy function to the advantage of all, particularly those who oppose such regulation of any type whatsoever? Where would they be and where would we all be

PROS AND CONS OF EQUALITY

without these programs and policies created to resolve inefficiencies and obstructions by power and privilege seekers who are willing to trample on equality of opportunity and deny the workers reasonable compensation for their willingness to do their part and support their families in a reasonable fashion?.

What determines power and privilege? What determines the level of income which a given profession receives for their contribution to the welfare of all? Who decides the hourly wage a common laborer receives? Who decides the hourly wage a skilled laborer receives? Who determines the wages or salary that a clerical worker receives? Who determines the salary that a manager or executive receives? Who determines the remuneration that a business owner receives?

Theoretically these are determined by supply, demand, special skills, special education, professional credentials, or personal relationships. In cases of unionization these wage and salary rates are negotiated between management and worker representatives. Salaries and wages paid in similarly related functions are used to exemplify standards of various industries of various skill levels. Often times these are determined by supply and demand, the number of workers seeking and the number of positions available. The only involvement of government regulations would be minimum wage laws in federal, state, of local jurisdictions.

Nicholas Goldberg asks the question in his article in the Los Angeles Times: "Why on Earth shouldn't we tax the super-rich more?" [69] He states that there has always been a troubling gap between rich and poor in this country and it is growing worse. He states that "billionaires with unimaginable fortunes go full years without paying any federal income taxes at all, while ordinary workers' wages have been stagnant for decades, and poor people are sleeping in miserable tent encampments on city streets." And he asks: "why are we so reluctant to take a reasonable share of Elon Musk's and Jeff Bezos' billions of dollars in taxes and put that money to work solving the country's enormous social problems." [69]

In California, with a population of about 40 million, more than 10 million people live close to the poverty line and most people believe that the American dream, that if you work hard you'll get ahead, is not true and probably never was. [69]

The records show that Jeff Bezos, George Soros, Michael Bloomberg, and Elon Musk, among our wealthiest billionaires, all have had years when they didn't pay any income tax at all. It has been reported that in 1950 the super-rich paid 70% of their income in taxes. [69]

We can assume that they didn't miss any meals by doing so. But, unfortunately for the country, by 2017 that rate had been reduced to 25% - more evidence that our financial system and the laws, rules, and regulations that administer it, are proposed, promoted, enacted, and enforced by those elected by the power of money through campaign finance, which assures their willingness to play by the rules dictated by those who own the gold. If they say it's not about money, it's about money!

CHAPTER **12**

What's it all About?

WHAT'S IT ALL about anyway? Kind of reminds us of the song "Alfie." "What's it all about Alfie, should we take more than we give, or should we be true Alfie?"

We are all in this thing called life together and we will all succeed or fail together regardless of how much we end up with as individuals. Is our society based upon "the one with the most toys wins?" What is wrong with wanting everyone to prosper at some level instead of the rich and powerful amassing huge fortunes and enormous collections of real property so efficiently that the working class either struggles constantly or fails miserably to compete for a comfortable life for themselves and their families, one with ample opportunities for advanced education and pursuit of goals of meaningful careers.

How much can one consume? How many homes can one utilize and enjoy? How many vehicles? How much stuff considered disposable is trashed instead of shared with those who can find remaining value? How much is our economy deterred by the stockpiling of money and possessions which are not needed or utilized, instead of the ownership or use being made available to a larger section of humanity and thereby enhancing economic opportunities?

What would be the benefit on the sustainability of planet earth for human habitation if the rich and powerful and others who earn more that they need for living comfortably would cease unnecessary travel by automobile, train, bus, and airplane all over the country and all over the world? A large percentage of trips are unnecessary and could be avoided without lessening the pleasure of life or the fulfilling of necessities. They are done for pleasure and impression of importance to others. The attitude is "look at me, how important I am."

Why do entertainers earn much more for their contribution to society than teachers? That is difficult to justify on the basis of a level of importance. Teaching is the most important profession for the future of humanity. Becoming a teacher requires more education than many other professions but in general provides limited income. Apparently, it is all based on supply and demand and potential opportunities. What are the chances of any one person becoming an actor making substantial income? What are the chances of someone on the legal profession making much greater income than others in the profession? What are the chances of a small business owner making income equivalent to a big business? All these questions rely on luck, being in the right place the right time, open doors, references, connections, luck of the draw. Can we make it fair? Do we want it all to be fair and equitable? Is there any opportunity to make it fair and equitable?

The only obvious method of leveling the playing field, as we discussed in a previous chapter is through fair and equitable taxation. Is that possible under our present laws, rules, and regulations? Of course it isn't. The tax laws under our system of government are enacted by those with the power of money because they finance the campaign expenses for those with the power of the vote. It's as simple as that. It doesn't require advanced education to understand this. It is common sense. It isn't even necessary to try to cover it up. It is entirely legal bribery.

In a campaign for election if one candidate offers you a hundred dollars for your campaign and another offers you ten thousand, who are you going to listen to? Who are you going to support? Who are you going to favor when decisions are made? These are rhetorical questions. The answers are obvious. And the answers confirm the allegations of many of our scholars in history, government, and

WHAT'S IT ALL ABOUT?

democracy – that our system of government is no longer, if it was ever, a democracy. A democracy is defined as a government of, by, and for the people, all the people, not just the rich and powerful – and for it to be of, by, and for the people, money must be eliminated as method of influencing voters.

Power and privilege should not be allowed to disenfranchise the powerless nor should it be allowed to provide inordinate economic advantage for those who have garnered fame and fortune. When the business owner or corporate executive earns more income or profits than the workers it should be balanced as a natural result of free-enterprise. When the difference is egregious and perhaps unreasonable, it should be challenged and regulated through taxation. Those who enjoy greater prosperity from an economic system than others should pay a proportionately higher portion of the cost of maintaining the system. Nothing could more fair and natural in a perceived democratic economy such as ours.

Jaron Lanier, in his book "Who Owns the Future," [70] offers a very prudent and penetrating analysis about economic stability and equal opportunity. He tells us that big business has changed. The internet has enabled companies to do billions in dollars of business with a few employees. Kodak once had 140,000 employees and was worth $28 billion. Today they are bankrupt. He tells us that Instagram was sold to Facebook for a billion dollars in 2012 and they had 13 employees. The seeking of online attention only turns into money for a token minority of ordinary people. In other words uncompensated sources, ordinary people, make networks valuable. Power and clout (i.e. power and privilege) are not equitably distributed. The very few prosper enormously from the efforts of the many and Lanier suggests that "monetizing the efforts from ordinary people will lead to a better future for all – as opposed to centralizing wealth and limiting overall economic growth." [70]

Lanier tells us that a market economy cannot thrive without the well-being of average people, without factories with multitudes of customers, without banks with multitudes of reliable businesses. He tells us that even the ultra-rich are best served by a healthy middle class, that a strong middle class does more to make a country stable and successful than anything else. [70]

The subject of power and privilege

The subject of power and privilege extends well beyond a struggle between individuals. Its reaches are worldwide between countries. The struggle for power and privilege, purportedly as an engine for prosperity, has a negative effect on economic progress as it feeds the huge division of income and wealth which serves as a deterrent to equality of opportunity and thus leads to an unproductive division of wealth. The large majority of wealth garnered by the rich and powerful leads to the majority of wealth lying idle in foreign and domestic bank accounts and investment in unproductive assets, while investment in job producing investments and additional income in the hands of the workers would have an immediate positive effect on the economy. Every dollar earned for low to mid wage earners would be plowed directly back into the economy for the necessities of life, such as food, clothing, housing, healthcare, education, and recreation. Any excess earnings for upper middle class wage or salaried workers would primarily be invested in improved housing, transportation, travel, or entertainment, rather than ending up in dormant savings accounts, thereby offering no economic impact.

Income received in excess of that mentioned above would be more likely to be considered non-productive capital enhancing the value of investment vehicles such as stocks and bonds – investment with little positive impact on the job-producing goods and services industries.

CHAPTER **13**

The Impact of the Cost of War

WHEN I THINK about war casualties it always reminds me of what Rodney King said after the police brutally beat him in 1992 during the Los Angeles riots: "People, can't we all get along?"

Well, the records before and since then would indicate very persuasively that we can't. According to the record books over the last 100 years the costs of war for the U.S. have exceeded $6 trillion. That amounts to about $60 billion annually. Total fatalities from these conflicts amount to more than a million.

Like former U.S. Senator Everett Dirksen is quoted as having said many years ago, "a million here and a million there, the first thing you know you're talking about real money." I guess we could upgrade that quote to current dollars and say, "a trillion here and a trillion there, the first thing you know you're talking about real money." And we could say, "a hundred thousand here and a hundred thousand there, the first thing you know you're talking about real people."

All this war, loss of life, and wasted resources may have been entirely necessary for the protection of our country and our citizenry. We uninformed bystanders certainly can't make educated judgements regarding the necessity or prudence of these decisions. However, there are some important steps we can take. We should insist upon a constitutional amendment requiring that a president cannot enter us into a conflict or war without the approval of Congress. And we should consider the question of whether we should require that any president entering us into warfare, whether it is called a police action or a war, must lead the troops into battle – not from the sidelines, not from the U. S. Capitol, but on the front lines of battle. Justification to send you or me or your kids or mine into battle should require the willingness of the person declaring war or armed conflict to lead the troops into battle.

Anything short of that leaves uncertainty of necessity and suspicion of history book legacy at the expense of those who fight the battles. And, furthermore, if such a policy were initiated by all countries, we may have seen our last war.

The following analysis of warfare reflects the documented human and financial costs of warfare. Undetermined is the immeasurable long term impact of these actual costs on the value of the contribution of those lost in these wars and the loss of the benefit of the funds expended which could have provided economic value elsewhere. And also undetermined is the mental and physical impact on those who fought the battles and their loved ones who suffered while they were gone to war or if they fail to return.

Costs of wars in lives and current dollars: [71]

Conflict	Duration	U.S. Fatalities	Dollars
American Civil War	1861-1865	620,000	$80 Million
World War I	1917-1918	116,000	$300 Billion
World War II	1939-1945	405,000	$4 Trillion
Korean War	1950-1953	36,000	$300 Billion
Vietnam War	1965-1973	58,000	$700 Billion
Persian Gulf War	1990-1991	300	$100 Billion
9/11 War on Terror	2001-2021	7,000	$1 Trillion
Total Costs	**160 years**	**1.25 Million**	**$6.4 Trillion**

THE IMPACT OF THE COST OF WAR

The imperativeness, urgency, decision making, or resultant value of the outcome of these wars and conflicts is subject matter which fills many shelves in public libraries. The opinions are based upon personal valuations, however, the total costs in lives and dollars and the impact on the Earth's resources are indeterminate. Military minded leaders may support a favorable valuation. Peace lovers may disagree. Those of us who lost loved ones may disagree. Those who experienced action and have mental or physical deficiencies therefrom, or their loved ones, may disagree. One can assume that those in positions of power made the decisions based upon their best judgement and the advice of other governmental leaders. Nevertheless, the quest for or the protection of power and privilege must have entered into the decision making process as well as the quest for reputation of leadership success or failure.

An example of presidential decisions and the factors that enter into them can be reflected in the situation of President Lyndon Johnson when he was reluctant to become involved in Vietnam in 1970. His statement:

> "I knew from the start that I was bound to be crucified either way I moved. If I left the woman I really loved – the Great Society – in order to get involved in that bitch of a war on the other side of the world, then I would lose everything at home… But if I left that war and let the Communists take over South Vietnam, then I would be seen as a coward and my nation would be seen as an appeaser and we would both find it impossible to accomplish anything for anybody anywhere on the entire globe."

We can read from his comment that the perception of the effect on the personal reputation of the decision maker can play a significant part in critical decisions made which effect those who confront the enemy, those who had no part in the decision making, but who are on the front line of battle. The soldiers don't get to vote on going to war. One could assume that if the leaders had to be soldiers we would have fewer wars. No one really knows the fear of bombs, guns, and bayonets until they experience it personally, on the battlefield with a gun in hand and their life at stake in at any moment. No one is more anxious to resolve a conflict without killing each other than those of the frontlines of battle at the command of leaders in the background who are protected from harm.

CHAPTER **14**

The Path to Power and Privilege

DOES WEALTH ENABLE power and privilege or does power and privilege enable creation of wealth? Well, I guess one could say it is like the question, "what came first, the chicken or the egg?"

Wealth is created either from earning extraordinary profits from financial endeavors, extraordinary remuneration for one's efforts in service of others, or by the good fortune of parents who have mastered the challenge of wealth accumulation.

Who owns the wealth?

As of Q1 of 2021, the top 10 percent of the wealthy held about 70 percent of total U.S. net worth, the value of all assets minus all liabilities. The top 1 percent held about 32 percent, while the next 9 percent held approximately another half at 38 percent. [72]

Concentration of Wealth Ownership [72]			
Wealth Class	Top 1%	Top 10%	Bottom 50%
Percent Owned	32.1%	69.8%	2%

It would be difficult to consider such a distribution of wealth ownership as indicative of an equitable system of free-enterprise. It may be more correctly interpreted as plutocracy, with wealth owned by the few who control the political system and thereby control the laws which regulate taxation which is the only method which could be productive in leveling the playing field of wealth distribution.

The distribution of wealth differs from the income distribution in that it looks at the economic distribution of ownership of the assets in a society, rather than the current income of members of that society. According to the International Association for Research in Income and Wealth, "the world distribution of wealth is much more unequal than that of income." [73]

There have been many different types of theories used to model aspects of the distribution and holdings of wealth. Before the 1960s, the data regarding this was collected mostly from wealth tax and estate tax records. The results from these sources tended to show that the distribution of wealth was very unequal, and that material inheritance had a big role in the matter of wealth differences and in the transmission of the status of wealth from generation to generation. There was also reason to believe that the inequality in wealth was shrinking over time. [73]

A study by the World Institute for Development Economics Research at United Nations University reports that the richest 1% of adults alone owned 40% of global assets in the year 2000, and that the richest 10% of adults accounted for 85% of the world total. The bottom half of the world adult population owned 1% of global wealth.

Concentration of Global Wealth Ownership [74]			
Wealth Class	**Top 1%**	**Top 10%**	**Bottom 50%**
Percent Owned	40%	85%	1%

Below is a comparison of the distribution of wealth in various nations in 2021. These are average wealth per adult comparisons, without comparisons of wealth distribution among income groups of each nation. (Credit Suisse Research Institute's "Global Wealth Databook", Table 3-1, published 2021.) [74]

Average Wealth Per Adult Comparison		
Country	Number of Adults	Wealth per Adult
United States	250,000,000	$505
Sweden	8,000,000	$328
United Kingdom	50,000,000	$305
Japan	100,000,000	$260
China	1,100,000,000	$68
Russia	112,000,000	$27
India	900,000,000	$14

Approximated for illustration purposes

Additional figures provided by Wikipedia: [75]

Wealth Owned by Income Groups			
Year	Top 1%	Top 10%	Bottom 50%
1989	30%		8%
2013		76%	1%
2021	32%	70%	2%

As one can see from these limited figures, we seem to have reached a plateau with the top ten percent of income earners owning about seventy-five percent of all wealth and the bottom fifty percent owning one or two percent of all wealth. This doesn't seem to be a financially healthy distribution of wealth. What is the impact of wealth distribution on the environment?

What about the waste of natural resources? What about global warming which has become the most important consideration we must address today? Perhaps we should question the impact on the atmosphere by those with excess wealth as they pursue personal pleasure without concern for their financial expenditures or their impact on the air we breathe and the heating of the planet. How much waste of valuable resources is expended each day or each year around the globe without regard for the planet's ability to sustain livable conditions for future generations of humans? That would be an important equation to pursue. We all have an impact on the environment with our lifestyles but the impact seems to follow a natural increase as income and wealth increase.

How much space must the wealthy heat, cool, and maintain to have a comfortable life? How many homes must they maintain? How many air flights all over the world must they make to enjoy their wealth? The difference in the footprint of a poor person, or a middle income person, or a wealthy person on the sustainability of humanity would notably be significantly different.

We should consider these effects of power and privilege on the environment, not for an exercise to place blame, but to highlight the steps necessary to sustain human life. Our continuous waste and deterioration of vital resources are mentioned in small print in the news media when it should be in bold caps. But, of course, we have acknowledged that the rich and powerful have no intention of exposing this in an alarming way. And, as we have pointed out, the rich and powerful own and control the media. It seems apparent that their lifestyles have a larger than normal impact on the environment.

One could question the lack of fiduciary responsibility of enabling and supporting a distribution of income and wealth which is a deterrent to the financial integrity of the economy. The benefit of more equality in income and wealth should be apparent to any student of finance or financial wizard. The value of more expendable funds entering into the flow of goods and services as opposed to lying dormant in the hands of the wealthy would assuredly provide

THE PATH TO POWER AND PRIVILEGE

greater economic value rather than enhancing the rewards of power and privilege and thereby enabling unnecessary environmental contamination for person pleasure.

Consider the progression of planetary atmospheric conditions over time directly affected by humanity.

When earth was first formed, its atmosphere was likely composed of hydrogen, helium, and other gases that contained hydrogen. Yet this atmosphere didn't last for very long because the solar wind from the sun blew it away. Solar wind is a stream of charged particles such as electrons, protons, and alpha particles. We now have a magnetic field surrounding earth that shields us from solar wind. [76]

The second atmosphere formed a little after 4.5 million years ago and was produced due to volcanic outgassing. Outgassing is the release of gas that was trapped in some other material. In this case, volcanic outgassing released hot gases trapped deep within the interior of the planet. Water vapor, carbon dioxide, methane, ammonia, and other gases similar to the ones produced by volcanoes today were expelled.[76]

Over a vast amount of time, millions of years, the earth gradually cooled. When the temperature dropped enough, water vapor condensed and went from a gas to liquid form. This created clouds. From these clouds, the oceans formed and the oceans absorbed a lot of the carbon dioxide in the atmosphere. A small amount of oxygen was produced by the photolysis of carbon dioxide and water vapor by ultraviolet radiation. [77]

Lastly, we have the third atmosphere. Around 2.5 million years ago the amount of oxygen available in the atmosphere started to rise due to the evolution of photosynthetic organisms that produced oxygen. These organisms were oceanic cyanobacteria. Over time, aerobic organisms evolved and consumed some of the oxygen produced. [78]

The present day atmosphere has arisen from long drawn out processes which have been taking place on Earth.

Earlier, there was an abundance of methane and nitrogen,

ammonia and almost nil percentage of oxygen.

But these conditions created life, which expanded and took in carbon compounds to give oxygen molecules or its compounds. About 3 billion years ago, these life forms which we often call plants, ruled the planet. By this time there was an abundance of oxygen, some CO^2 or carbon compounds, and huge amounts of nitrogen. [78]

The proportion of oxygen went up because of photosynthesis by plants. The proportion of carbon dioxide went down because it was locked up in sedimentary rocks (such as limestone) and in fossil fuels, it was absorbed by plants for photosynthesis, and it dissolved in the oceans.

The burning of fossil fuels is adding carbon dioxide to the atmosphere faster than it can be removed, thus the level of carbon dioxide in the atmosphere is increasing.

Global atmospheric carbon dioxide concentration has increased from 325 parts per million in 1970 to 410 parts per million in 2019. [78]

These facts and figures are included not as a lesson of planetary science or a need to vocalize them in ordinary conversation. They are included to emphasize the fragility of our atmosphere over time and the adverse effects of human pollution which increases as population increases and the continuous progression of comfort and pleasure are pursued and enjoyed. It kind of warns us again of the cliché of the medical world, one which we should be constantly cognizant of: "the operation was a success but the patient died."

CHAPTER **15**

Political Opinions

SIMPLE CLICHÉS, SOME made in humor, some made to simplify a point, some merely as cleverness, can often say more in fewer words and provide a lasting conclusion or opinion.

Napoleon, the French military leader and emperor who conquered much of Europe in the early 19th century:

"Religion is what keeps the poor from murdering the rich."

Honore' de Balzac, French literary artist who produced a vast number of novels and short stories collectively called La Comédie Humaine (The Human Comedy).:

"Behind every great fortune there is a crime."

George Bernard Shaw, an Irish playwright and political activist. His influence on Western theatre, culture and politics extended from the 1880s to his death and beyond:

"A government that robs Peter to pay Paul can always depend on the support of Paul."

Lily Tomlin, an American actress, comedian, writer, singer and producer:

"The trouble with a rat race is that even if you win you are still a rat."

Ray Charles in the song: "Them That's Got" –

"Them that's got is them that gets, and I ain't got nothin yet."

Kurt Vonnegut, "Timequake"

Vonnegut speaking about prohibition in 1919:

"Intoxicating liquors did not become lawful again until 1933. By then, the bootlegger, Al Capone, owned Chicago, and Joseph Kennedy was a multimillionaire."

Vonnegut speaking of fiduciaries:

"We have faithless custodians of capital making themselves millionaires and billionaires while playing beanbag with money better spent on creating meaningful jobs and training people to fill them, and raising our young and retiring our old in surroundings of respect and safety."

The political stances of these well-known commentators mentioned above, some quoted from a century or more ago, still ring true today. The same division remains, with Democrats supporting the right to vote for every citizen and encouraging everyone to vote, while Republicans attempt to disenfranchise as many voters as possible by limiting access to the ballot box, limiting the hours available for voting, or challenging the legality of those who are attempting to exercise their right to vote.

POLITICAL OPINIONS

Republicans have apparently come to realize that the more people that vote, the less their chance of winning. So, if you can't win with your principles, make it difficult for those who oppose your viewpoints to vote and tilt the playing field in your favor. This may be considered demagoguery or disenfranchisement, or it may be considered fair game, but it surely indicates a willingness to deny democratic principles by any measure available in controlling the electoral process and discouraging those who are merely exercising their right and obligation as citizens to an undeterred access to the ballot box. It says that the only thing that matters to some is winning and maintaining power.

Fame and fortune, fortune and fame, which comes first, the chicken or the egg? Well, both are in correct order in some situations. Eggs make chickens and chickens make eggs. The same can be said for fame and fortune. Some attain fame by first attaining fortune and some attain fortune by first attaining fame. And, of course, some attain both by careful selection of the proper parents – birthright.

Making the case for eliminating money from political campaigns can be argued by this principle. What comes first the political contributions or the principles of government? Not to say that there aren't many honest and dedicated officials who serve humanity in various roles, but the influence of money cannot be denied. As we queried earlier, who gets through to the elected official, you or major campaign contributors. The chicken or the egg analogy may apply here. The political contributions breed the chicken which lays the golden egg.

Where do we go from here? What steps must be taken to control our incessant drive for fame, fortune, comfort, and pleasure? What steps must be taken to instill a reduction or elimination of the increasing population and thus the ever growing consumption of finite natural resources and the ever growing emission of gases which continue to consolidate in a layer of human-generated waste which surrounds the planet and allows the heat of the sun to enter the Earth's atmosphere and prevents it from escaping out of our atmosphere.

It seems that the quest for and distribution of income and wealth must be a source of the dilemma. It seems that the Supreme Court decision which allows corporations to finance political campaigns is a source of the dilemma.

The inequality of the distribution of income has exacerbated the inequality in the distribution of wealth. This inequality can be partially corrected by our system of taxation. This will require the populous to gain control of the government by wresting it from the domination of the rich and powerful. The financing of political campaigns is the obvious cause of our inability to legislate effective change in taxation. We must have a collective uprising of the people. It must happen at the ballot box. There is no other way. Such change will require a broad sense of urgency and the willingness to support and participate in a massive movement of the people. Funding must come from the masses, perhaps by diversion of current tax laws and redirection of taxes to the control of a movement to gain control by the people, the condition which was anticipated by the populous at the inception of our quasi-democracy, not by those who initiated our independence from foreign control, but by the populous.

CHAPTER **16**

Measures for Enhancing Equality

MANY SUGGESTIONS BY scholars, students, and political activists have been offered without being implemented as potential corrective measures. And many of them seem worthy of implementation in our system of income and taxation. Political persuasions facilitated by our undemocratic campaign finance practices have prevented potentially promising changes to level the field from becoming lawful. The pros and cons of such changes have prevented implementation of such measures to improve equality of opportunity in distribution of income and wealth.

Ban political campaign contributions.

Perhaps political candidates should be elected by popular vote without the advantage or disadvantage of the power of money in campaign finance. Every vote could be based upon the decision of the voter from information provided by the candidate of choice. Allowing the power of money to influence the voter merely enables the rich and powerful to control governmental actions to their advantage. It is a deterrent for those without financial advantage, thereby allowing money to vote instead of assuring voter equality. Free speech gets in the way.

Scale income taxation rates on ability to pay.

The rate of taxation should be based upon a subsistence level of income under which no tax would be due and an increasing level of taxation based upon established levels of income above subsistence levels. Rates could be scaled from a minimum rate of ten percent of gross income and a maximum rate of sixty percent of gross income, or such other scale which is deemed appropriate. Campaign contributions get in the way.

Disallow deductions of any kind from taxable income.

Taxation should be based upon gross income, thereby eliminating deductions of any type. This would eliminate questionable or unverifiable deductions and deductions which are not available for low wage earners. It would also provide a significant reduction in the cost of income tax return review by the Internal Revenue Service. The rich and powerful get in the way.

Finance political campaigns by government grants.

To provide a level playing in the electoral process there should be equality of funding for campaign expenditures. Otherwise, those with greater income or wealth can continue to control government regulations without restraint, such as our current system provides. Any candidate or supporter of any candidate who provides financing for a candidate should be required to provide equal funds for all candidates. Money should no longer be allowed to vote. Power and privilege get in the way.

Define Free Speech to mean oral or written, not money.

The Supreme Court decided in the Citizens United decision that corporations could vote in political campaigns with their power of money. Money was declared to be considered as Free Speech. This policy should be eliminated entirely by amending the constitution to disallow unequal funding of candidates for office. The conservative Supreme Court gets in the way.

Elect Supreme Court Justices by popular vote.

Supreme Court justices should be elected by popular vote. Under our present process Supreme Court justices are appointed by the U.S. president and approved by Congress. Under this process one political party who maintains a majority in Congress and the presidency can use their current political power to affect our laws, rules, and regulations for many years to come. The challenge of constitutional change gets in the way.

Allow only one term for Supreme Court Justices.

Supreme Court justices now have a lifetime appointment, an undemocratic form of government. They should be elected by popular vote and serve for a specific term without any extensions. The power of the judicial system gets in the way.

Allow only one term for all elected officials and court judges.

All elected officials and court judges should only serve one term. The voters should decide who serves and have the right to elect whomever they feel is deserving of the role and has the appropriate background and experience to serve in a professional and unbiased manner. Career officials, those serving continuously in one or more elective office become indebted to fund-raisers and campaign contributors. They then declare that campaign funders do not receive any favorable legislation or enforcement, but we know better. Human nature gets in the way. The only method of control is to eliminate campaign contributions. This may require electing those to office who will change the process of appointing judgeships who then favor those who appointed them. All elected officials who enjoy the financial rewards and power and privilege get in the way.

Provide reviews for government employment.

All candidates for government employment should be subject to an extensive review of their background and qualifications based upon those specified as necessary to perform the position sought. Free speech and equality of opportunity get in the way.

Provide reviews for continuation of government employment.

Government employees should be subject to periodic reviews to determine job performance by specially qualified staff members. Shortcomings determined should be detailed to provide the given employee ample opportunity of correction or improvement. All government employees get in the way.

Apportion U.S. senators to each state based upon population.

U.S. senators are now apportioned at two per state, a policy created at the formation of our nation to provide agreement of the smaller states to agree to the founding principles. This was not considered a major source of unequal rights at the time. Now it creates an undemocratic system of government. For example, in California the two U.S senators proportionally represent about twenty million people each. In much smaller population states the number each senator proportionally represents would be much smaller, a fraction of the number in California. The smaller states prevent the possibility of a constitutional amendment.

Require news publications to provide unbiased coverage.

Print publications and broadcast publications should be limited in their influence in the electoral process. There should be requirements to provide equivalent coverage of editorial content supporting candidates. Campaign contributions should be limited and monitored to provide equality of information about the candidates and their positions on important issues of governance. Free speech and the free press get in the way.

Assurance of a quasi-democratic government which we strive for will be difficult to establish without some meaningful changes to the U.S. Constitution. The present Supreme Court is considered by many of us to be conservative politically, a condition established by the appointment of conservative justices of like-minded political persuasions.

Although Supreme Court Chief Justice Roberts recently stated something to the effect that we don't have conservative justices or liberal justices, we just have justices - we know better. Supreme Court Justices have important biases regarding who rules the nation and how it is ruled. They certainly have their foot on the pedal of their own political persuasion. Those with the power to appoint justices appoint those who favor their political leanings. To establish a government which assures equality of opportunity in human rights of education, employment, health, and welfare we must make it clearly defined in such a way that eliminates interpretation based on the political beliefs of those chosen to interpret the issues, but according to more strict and defined constitutional provisions.

CHAPTER 17

Democratic Governments

The 10 most democratic nations in the world (2020): [79]

1. Norway
2. Iceland
3. Sweden
4. New Zealand
5. Finland
6. Ireland
7. Canada
8. Denmark
9. Australia
10. Taiwan

The United States didn't make the top twenty of countries considered as governed by some form of democracy. We were ranked as number twenty-one. We consider our country a democracy but we are more accurately defined as a democratic republic. Actually, our Pledge of Allegiance to the Flag of the United States of America states: "… and to the republic for which it stands." It doesn't mention democracy.

What is a democracy?

At its most fundamental, a democracy is a form of government in which a nation's citizens have the power to decide the laws under which they will live.

A dictionary definition: A government by the people; a form of government in which the supreme power is vested in the people and exercised directly by them or by their elected agents under a free electoral system.

For example, the United States is a representative democracy because most decisions are made not by the people themselves, but by representatives who act on the people's behalf. It is also an electoral democracy because those representatives are selected in elections. It is a presidential democracy because the head of government is also the head of state and leader of the executive branch. It is a constitutional democracy because its fundamental principles and laws are guided by a constitution. Some argue that this makes the U.S. a republic rather than a democracy.

There are multiple theories about what specific elements are required for a government to qualify as a democracy. For example, in preparing its annual Democracy Index, the Economist Intelligence Unit scores each of the world's countries in five distinct categories—which we can examine to determine several of the Economist's democratic wish list:[79]

1. A pluralistic system in which at least two legitimate-but-different political parties coexist
2. A free and fair electoral process that enables the people to choose between candidates from those parties
3. A government that operates openly and transparently, works for the good of all the people, respects its own rules, has proper checks and balances, and gives its citizens free choice and control over their lives
4. Politically engaged citizens who support democratic principles, "fight fair", vote regularly, accept the will of the voters,

and commit to a peaceful transfer of power after each election
5. An emphasis on preserving civil liberties and personal freedoms of both the majority and minorities
6. A free and independent media unhindered by government interference, influence, or intimidation

Stanford University political scientist Larry Diamond has a similar list, maintaining that any democracy must include four key elements: [79]

1. A political system for choosing and replacing the government through free and fair elections
2. The active participation of the people, as citizens, in politics and civic life
3. Protection of the human rights of all citizens
4. A rule of law in which the laws and procedures apply equally to all citizens

The principles and intentions of democracies are common and universally accepted in theory. It is in application and practice that they vary and become difficult to define as democratic. And, as we all know, it is politics that gets in the way. Obviously, everyone doesn't think the same way about what constitutes a democracy. But a more challenging issue preventing a democracy to work effectively is always power and privilege. As someone stated, "power corrupts and absolute power corrupts absolutely." As we review the history of democracies this becomes apparent. Even under seemingly stringent rules of governance people don't think the same way and the exercise of power gets in the way.

The Costs of Democracy

And, back to the same issue which causes potential democracies to become quasi-democracies or plutocracies – campaign finance - allegiance to those who provide it, promises to those who dangle it for support, fulfillment to assure continuity of financial support. In

DEMOCRATIC GOVERNMENTS

other words, money gets in the way.

Money drives ambition. It drives choices. It drives efforts. It drives education. It drives friendships. It drives disputes. It drives class warfare. It drives crime. It drives our rules, laws, ordinances, and interpretations of same. It drives our judicial system. It drives our penal system. It engenders a repeat of the oft-quoted phrase: "if they say it's not about money, it's about money."

If we want to simplify our system of exchange maybe we should return to the days of trading a chicken for a loaf of bread. This would eliminate the enormous cost to society of financing our legal system, a cost which is probably impossible to tabulate. What is the total cost of our legal system in a democracy? One would have to include the entire cost of the judicial system – thousands of judges, thousands of juries, hundreds of court houses, hundreds of court rooms, millions of lawyers, millions of textbooks, hundreds of textbook writers and publishers, thousands of professors, hundreds of office buildings, hundreds of jails and prisons, millions of law enforcement personnel and on and on.

Calculating the cost of all this is only half of the equation. What about the lack of efforts and functions of all those associated with law enforcement and the other aspects of our legal system in more useful functions such as providing the sustenance we all need such as food, shelter, clothing, education, entertainment, health services, and other essential necessities of humanity?

The Cost of the U.S. Criminal Justice System [80]

The direct governmental cost of our corrections and criminal justice system was $295.6 billion in 2016, according to the Bureau of Justice Statistics. With more than 2.2 million people incarcerated, this sum amounts to nearly $134,400 per person detained.

The United States spends nearly $300 billion annually to police communities and incarcerate 2.2 million people. [81]

The societal costs of incarceration—lost earnings, adverse health effects, and the damage to the families of the incarcerated—are estimated at up to three times the direct costs, bringing the total burden of our criminal justice system to $1.2 trillion.

The outcomes of this expense are only a marginal reduction in crime, reduced earnings for the convicted, and a high likelihood of formerly incarcerated individuals returning to prison.

A criminal justice system is vital to ensuring laws are obeyed, the public is safe, and rights are protected. Key elements of such a system include incapacitating people who have broken the law, deterring others from doing the same, and rehabilitating offenders to prevent reoccurrence. A fair and just system must provide due process, protect the rights of the innocent, and provide those protections equally to all people. Further, victims of crimes should be compensated for their sufferings and made whole, insofar as it is possible. To the extent these goals are achieved, such outcomes are the benefits of a robust criminal justice system and an indication of its effectiveness. The resources employed to achieve those outcomes, as well as any errors and collateral damage caused in the pursuit of justice, are the costs.

A well-functioning criminal justice system may exhibit low or falling crime rates, low recidivism rates, and the ability to move on with one's life after a person's sentence has been served or debt paid, as well as the ability of victims to be compensated for the wrongs committed against them. But the value of these attributes is subjective and will differ from individual to individual based on a personal evaluation of safety, life, and property.

The direct governmental cost of our corrections and criminal justice system was $295.6 billion in 2016, according to the Bureau of Justice Statistics. With more than 2.2 million people incarcerated, this sum amounts to nearly $134,400 per person detained.

Roughly half of these funds—$142.5 billion—are dedicated to police protection. The next largest share of this expense—$88.5 billion—is the cost of operating the nation's prisons, jails, and parole and probation systems. The remainder—$64.7 billion—is spent on the judicial and legal systems. As shown in the following chart, local governments pay more than half of the total costs—mostly for policing, while the federal government pays just one-sixth. States spend the most on corrections, a reflection of the fact that nearly 60 percent of all detainees (1.3 million people) are held in state prisons.

A study from Washington University in St. Louis estimates that the broader societal costs put the total burden at nearly $1.2 trillion, after accounting for consequences such as foregone wages, adverse health effects, and the detrimental effects on the children of incarcerated parents, as detailed below. Other studies have noted similar indirect costs. [81]

Paul Krugman expounds on the subject of economic measures to improve the economy. His suggestion is supported in his article: "Tax the Rich, Help America's Children." He discusses the pending legislation by the Biden administration to spend $3.5 trillion for important economic measures to be supported by taxes of incomes of billionaires and minimum taxes on corporations. [82]

Krugman mentions that despite 40 years of anti-tax propaganda that large majorities, including many Republicans, support higher taxes on corporations and the rich. He states that, " ... while Republican politicians, claiming that Democrats are anti-American and that Democratic proposals are Marxist, history tells us that the key elements of the legislation which would aid the middle-class and poor children together with higher taxes on the wealthy are quintessentially American ideas." [82]

He tells us that although the modern Republican Party is utterly committed to the proposition that low taxes on corporations and the rich are the key to economic success, there is no evidence that this is true." He states that: "... the historical correlation runs the other way" – and that, "the U.S. economy grew faster during periods when taxes at the top were relatively high than it did when they were low."

Krugman tells us that: "... the nation invented progressive taxation with the progressive income taxes and estate taxes on large incomes and estates since 1916." He quotes Theodore Roosevelt stating in 1906 that "it was essential to prevent the inheritance or transmission in their entirety of fortunes swollen beyond all healthy limits."

He called for a "heavy progressive tax on estates." [82]

This policy was established as a provision for retention of equality of opportunity more than a hundred years ago. Although the policy has perhaps been altered in enforcement, perhaps ignored, and perhaps violated by some for these hundred years, it is still in force and has never been successfully altered or declared unenforceable by Congress, nor declared as abusive or unfair as a progressive taxation policy by a majority of American taxpayers. It seems that a taxation policy based upon one's ability to pay has passed the test of time.

The common belief among our leaders past and present supports the fact that billionaires got to be billionaires by clever use of the market for goods and services which our economy provides, or by good fortune. There is little expression or argument by economists to support any reduction or elimination of progressive taxation, our only measure of protecting free enterprise as a viable economic system which enables fair trade and assures minimal living standards, without which billionaires would not exist.

This writing supports the belief that progressive taxation is not progressive enough. No one needs a billion dollars. No one can spend a billion dollars. One can lose a billion dollars on bad investments, but it would be unlikely that one could spend it on lavish living expenses or enjoyment.

Some of us believe that any accumulation of wealth by an individual in excess of a few million dollars should be taxed at a very high rate, a rate in excess of fifty percent, perhaps seventy-five percent. The governmental rules, laws, and regulations, along with the administration and enforcement of same, should be funded by excess wealth of individuals and corporations, not by the working-class who lack funding to support governmental influence in such matters. Again, who makes campaign contributions? Who owns the representatives who claim to be of, by, and for the people, but who are in large part of, by, and for the corporations and corporate owners who prosper from the protections provided by such legislation and enforcement?

Corporations should be taxed much more progressively. The rules

DEMOCRATIC GOVERNMENTS

and regulations which are essential in assuring an equitable economy have established the ability of corporations to protect shareholders from personal liabilities, the primary purpose of incorporation. This should not limit a corporations' duty to pay the cost of providing and protecting such privileges. Corporate owners would prefer that individual taxpayers pay for administration of these provisions. This cost should come from corporate profits.

Perhaps corporate profits should be regulated to shareholders on a limited basis of an established percentage of return on shareholder value, with all additional profits returned to the government through taxation to reimburse governmental expenses of administration and protections under the law, services without which corporate profits may be difficult to maintain.

Furthermore, corporate profits should be redefined by establishing a standard form of profit and loss statements. Depreciation of capital expenses should be eliminated as a deductible expense. These are investments. Capitalization of investments should not deductible as an expense. Their value typically does not diminish by usage, it continues year after year as long as the corporation continues to exist. Corporate profits should be determined by total income less actual out-of-pocket expenses, such as facilities, material and labor costs. Investments should not be recoverable from the citizens of the country by being deducted from corporate profits. By allowing the deduction of depreciation of capital investments, all of the citizens of the nation pay for these assets but get no benefit therefrom through taxation, actually just the opposite.

If you or I buy an automobile to drive to and from work do we get a tax deduction? If we rent a house or apartment do we get tax deduction? If we hire someone to perform work on the house or mow the lawn, do we get a tax deduction? No, we don't get tax deductions for any of these necessities of life. We pay tax on our gross income less the deductions provided by the individual tax code, which are limited to excess medical expenses, educational interest expenses and a few others. All legal deductions from personal income are listed in a previous chapter. They don't include deductions based on depreciation of the value of our automobiles or other necessary possessions. Only corporations get these deductions, not individuals. Decline of asset values should not be a deductible expense to reduce taxable income.

It seems that the only way to make taxation more fair and equitable would be to base it on gross income of individuals, corporations, limited liability companies, and any other legal entity. This could be structured to protect a sufficient level of income for individuals, families, proprietorships, partnerships, limited liability companies, and corporations. It would simplify the tax codes, the tax return review process, and the enforcement process. It would provide substantial savings of the cost of the Internal Revenue Service review and the enforcement agencies necessitated by our current system.

The likelihood of this coming about would again fall under the percentages of slim and none, simply because those who own the gold make the rules. It all cycles back to campaign finance. What are the chances of enacting new legislation that would provide more equality of income and wealth if those who enjoy inordinate income and wealth, either individually or as shareholders of corporations, can finance the campaigns of those who make the rules about taxation? This is obviously another rhetorical question, one which answers itself. The chances again are two - slim and none! Only the working-class vote against their own best interest - not intentionally, but because they are misled by messages afforded by campaign finance, a game in which they are ill-equipped to engage.

The lack of equal access to the hearts and minds of the populace due to the cost of the use of effective communication supports an unfortunate outcome. The important lessons of humanity, the real wisdom which is most important for assuring the endurance of the species, seem to only register in the forefront of our minds after we have struggled through the educational and employment years of our lives.

The young people of today must endure the results of the damage

DEMOCRATIC GOVERNMENTS

we have done to the life sustenance that nature provides. We should feel compelled to sustain the ability of the planet to support life for the benefit of future generations, not just the next generation or next few generations, but for millions of years to come. Without such a commitment the sustainability of our climate will continue to diminish over each century until the planet will no longer support human life. You and I won't see that outcome, but millions will if we don't do the right things for lessening the effects of climate change sooner rather later. The chances of that don't look favorable. I hope I'm wrong.

Speaking of history: "The great historian Edward Gibbon once called history: "Little more than the register of the crimes, follies, and misfortunes of mankind." Voltaire called it a trick played by the living upon the dead. Thomas Carlyle said it was "a distillation of rumor." Henry Ford said history is "more or less bunk." ("Don't Know Much About History," [83]

Kenneth C. Davis tells us in his closing line of "Don't Know Much About History," "….. remembering America's history becomes all the more important. What's past, after all, is prologue. [83]

Senator Joe Manchin of West Virginia tries to influence the history books as he states that he is against taxing the rich because they have "contributed to society and create lots of jobs and invest lots of money." And he maintains that stance although the gap between the rich and poor here is larger than any other comparable nation. And he ignores the fact that the rich got there wealth by creating lots of jobs and investing lots of money they earned on the backs of underpaid and under-protected workers in the coal mines of West Virginia. He is just another self-serving hypocrite who gained fame by gaining fortunes on the backs of the workers. It is doubtful that he could state a scholastically acceptable definition of democracy but as the member of a family of wealth from coal mining he surely has his self-serving opinions about taxation.

Nicholas Goldberg, writing in the L.A. Times, tells us that "billionaires with unimaginable and un-spendable fortunes go full years without paying any federal income taxes. Meanwhile, ordinary workers' wages have been stagnant for decades, and poor people are sleeping in miserable tent encampments on city streets." He asks: "Why are we so reluctant to take a reasonable share of Elon Musk's and Jeff Bezos' billions of dollars in taxes and put that money to work solving the country's enormous social problems?" [69]

Goldberg states that realizing that "the notion that with a little luck, that any citizen can become the next John D. Rockefeller, Warren Buffet, or Bill Gates is a fairy tale." He states that, "while homelessness has reached crisis proportions in cities including Los Angeles, San Francisco, and New York, Elon Musk can lose $50 billion when Tesla stock falls and still be worth $200 billion." He tells us that former Labor Secretary Robert Reich calculated that it would take the median U.S. worker over 4 million years to earn that much money." He tells us that in California more than one-third of the state's nearly 40 million people live close to the poverty line while Forbes counts 189 billionaires living in the state." [69]

And in spite of the obvious fact that the distribution of income and wealth has reached unprecedented and economically critical proportions, former president Trump boasts of his important accomplishment of getting tax reductions for the wealthy. We obviously wasted, and were damaged by, the four years of Donald Trump as president. Now Joe Biden must make the right moves to right these wrongs and provide an economy that works for all Americans, not just the privileged few.

Those of us who realize that our tax laws are unfair to the wage-earners and favorable to the rich and powerful now anxiously await an answer to our call for a more just system based upon the ability to pay and a redistribution of inordinate wealth to the benefit of all citizens. Those who prosper the most from our system of free enterprise should pay a much larger portion of their income and wealth to support an ongoing stable economy. Without such a system of leveling the income and wealth we cannot maintain an equitable economy to sustain consistent growth by providing a reasonable standard of living, education, and healthcare for all citizens. In other words, we can't have a vibrant economy if the workers don't have money in their pockets. The closer we can get to a fair and equitable system of taxation, the better our chances are of being a democracy, that cherished form of government which has been evasive due to the political control of our system of law, rules, and regulation by the rich and powerful by way of campaign finance. Without changing campaign finance we cannot achieve a more democratically balanced economy with lasting sustainability.

To achieve a more democratic form of government we must first change our electoral process from one which is controlled by the power of money to one which is controlled by all the voters. This simply cannot be accomplished unless campaign finance is eliminated as a method of inordinately influencing the voters to vote against their own best interest. We must finance political campaigns exclusively by public funds equitably disbursed to eligible candidates for federal offices both presidential and congressional. Until that is achieved we will struggle to maintain an economy which enables a sustainable measure of equality of opportunity.

One method of financing campaigns would be taxing the wealth of the rich and powerful individuals and corporations which they have cleverly garnered by creating and enforcing laws and rules of taxation which enabled such accumulation of excess wealth.

CHAPTER **18**

Greed Revisited

AS WAS ILLUSTRATED in great detail in "Greed Disease," a 2017 publication by this author, the effects of greed on society place an insurmountable burden on the working class, those who do what is necessary to provide the best standard of living possible under a system of inequality of opportunity, a condition which is beyond their control.

This kind of reminds one of what Kermit the Frog said on Sesame Street, "It isn't easy being green." The working class of society could likewise state their challenge regarding the inexorable economic and political discrimination, "It isn't easy being a working-class consumer." But, those who we consider members of the working class don't have time to sing a song of protest. They are too busy trying to provide ample food, shelter, clothing, healthcare, education, transportation, and entertainment for themselves and their families, the costs of which are established and maintained by those making the rules of the game, the rich and powerful. It's like a poker game with a stacked deck. This challenge has changed the formerly desirable family functions of the working-father and the stay-at-home mother. Now both parents are employed in most families and childcare has become commonplace as a major budgetary obligation.

Not to say that this isn't a challenge for everyone, but the impact is not universal. It depends somewhat upon the power and privilege which one attains or lacks due to birthright, family connections, cleverness, or good or bad fortunes in life.

In the book "Greed Disease" the list of CEO salaries was compared to the average worker. The comparisons were alarming to say the least. The list of the highest paid CEOs' annual salaries in 2015 was noted as being between $18 million and $143 million, which would typically be $10,000 to $70,000 per hour. The average worker remuneration in 2015 was about $48,000 annually or $4,000 per month, about $20.00 per hour. The federal minimum wage is $7.25 per hour, which would be about $15,000 annually or $1,250 per month, an income level which would be unlikely to support an individual, let alone a family, with any sense of a comfortable standard of living or equality of opportunity.

Comparison of these levels of the salaries noted on the report reveals the highest paid CEO is paid about 3,000 times what the average worker is paid and the lowest paid CEO on the list was paid 400 times what the average employee was paid. It also revealed that in 2015 the average worker was paid $24.00 per hour, the lowest paid of the top 100 CEOs was paid $9,500 per hour, the average CEO was paid $40,500 per hour, and the highest paid CEO was paid $71,500 per hour. The fact that the CEO is worth from 1,000 to 4,000 times more to the given company than the average employee seems hard to justify. It is also noted that in 2017 the CEOs of large firms got the biggest raises in the last three years while the minimum wage remained at $7.25 per hour. The minimum wage set in 2009 has remained the same to the date of this publication, more than twelve years, while all of the above mentioned income increases were bestowed on the grateful recipients.

Inequality of income took a turn in favor of workers somewhat many years ago with the inception of unionization. Wages, hours, and working conditions were negotiated and administered by union organizations with the power of the labor strike as a negotiating tool of persuasion. This condition is still a factor in some employment categories, such as automobile manufacturing and perhaps a few others, but the strength of the union movement was hampered in large part by right-to-work laws and union-busting provisions enacted by the impact of corporate lobbyists who enabled union-busting activities to be rendered ineffective. This relates to our discussion about corporations having been granted the right to vote with their money in political campaigns – a lasting discrimination against the worker in favor of management. This reminds us of the old cliché "money talks and you-know-what walks."

Apparently some companies realize that a living wage for their employees is not only the best policy for the employees but a good policy for the company. As was cited previously, most economists and corporate executives seem to agree that the economy doesn't work well unless the workers have money in their pockets. Otherwise, who will buy the goods and services offered by entrepreneurs and corporations?

Greed is certainly a factor in business operations in general, perhaps rightly so, but goods and services must be priced commensurately with their value as determined by the user. The cost of goods or services is certainly a factor in pricing, but the value determined in the mind of the consumer with limited financial resources is a major factor in price consideration and one ignored at the seller's peril.

Pricing of consumer products is almost an economic science. It is based upon many factors other than the cost of providing the goods or services, plus a reasonable margin of profit. Pricing is sometimes based on cost, sometimes based on supply and demand, and often is based on what the traffic will bear. The pharmaceutical industry taught us a painful lesson in that science with their egregious pricing of many multiples of cost, all with an impact on the uninsured consumer or the costs of health insurance for the insured consumer. Cost seems to have no factor in pharmaceutical pricing. It is more focused on what the traffic will bear. This is enabled by elimination of competition created by legal methods such as branding, patenting, and other exclusionary tools of monopolization. Do the pharmacy companies care whether you live or die? We, of course, assume that they do care, but their pricing policies seem to indicate that they won't let that possibility interfere with making enormous profits for the benefit of corporate executives and shareholders.

This seemingly uncontrollable policy of pharmaceutical pricing lends additional strength to the calls for price controls of essential health products and practices. It is indeed a call for universal healthcare for all Americans, regardless of financial stature or perceived importance in societal hierarchy.

In Greed Disease we discussed the "Master Deceivers" who were exposed as exemplary of self-serving fraudulent activities. Enron, a major investor in the natural gas industry, was cited as an example corporate greed in creating deceptive trading of futures in the market to the tune of hundreds of billions of dollars. They hid financial losses by deceptive practices called 'mark to market" which overvalued future gas prices with nearly $350 billion in trades. The outcome resulted in Enron's collapse which affected the lives of thousands of employees. Their share price had risen to $90 or more and fell to less than $1.00 after they crashed. The three executives charged with responsibility for the fraudulent scheme, Andrew Fastow, Kenneth Lay and Jeffrey Skilling were convicted of fraud and conspiracy. Fastow and Skilling received long prison sentences and Lay died before being sentenced.

Waste Management Inc. was a big player in managing environmental engineering in the protection of groundwater. They launched a massive financial fraud in the 1990s by creating financial misstatements to meet predetermined earnings targets. The officers charged settled with the federal government for more than $30 million.

And, of course, our former president, Donald Trump, was forced to settle with victims of the Trump University scandal in 2017 whereby it claimed to help consumers make money in real estate by attending seminars, a claim determined to be a scam. Eric Schneiderman, New York Attorney General, considered Trump's actions the "hallmark of snake-oil salesmen." Trump settled the lawsuit for $25 million.

Trump has had a string of bankruptcies causing substantial losses for investors. These include the Trump Taj Majal, which was in debt for $3 billion of which Trump owed $900 million in personal liabilities. He made settlement by giving up some ownership share, sold his Trump shuttle airline and a 220-foot yacht.

Two years later he filed for Chapter 11 protection with the Atlantic City hotel-casinos, the Trump Plaza and Trump's Castle. Then a couple of years later similar results transpired which resulted in his resignation as chairman from Trump Hotels and Casino Resorts. The Trump Plaza casino was ultimately closed.

Trump's other business failures included Trump Steaks, Trump Airlines, Trump Vodka, Trump Mortgage, Trump Magazine, Trump University, and Trump Network.

Then we elected him president of the United States. And, after the voters rejected him for reelection, he still remains the uncrowned leader of the Republican Party. Go figure!

Corruption has not been limited to dishonest business would-be tycoons like Donald Trump. It also includes some of our elected and appointed government officials. It seems easy money lures many to end careers and reputations due to overzealous greed:

Chaka Fattah, a congressman, was convicted on twenty-three counts of racketeering, fraud, and other corruption charges. He was found guilty of misspending government funds and charity money to fund his campaign and personal expenses. He repaid a portion of an illegal loan from a wealthy friend with government funds and was sentenced to ten years in prison for money laundering, bribery, and fraud.

Rich Renzi, a congressman, was indicted for conspiracy, wire fraud, money laundering, extortion, and insurance fraud. He was charged with willfully embezzling from a risk retention company. He was convicted on 17 counts of personal gain and looting a family insurance business to pay for his campaign and was sentenced to three years in prison.

Duke Cunningham, a congressman, pleaded guilty of accepting $2.4 million in bribes and under-reporting his taxable income.

He pleaded guilty to conspiracy to commit bribery, mail fraud, wire fraud, and tax evasion. He was sentenced to eight years in prison and ordered to repay $1.8 million in restitution.

Frank Balance, a congressman, was indicted of federal charges including money laundering, mail fraud, and tax evasion. He was disbarred from the practice of law and sentenced to four years in prison.

Jim Traficant, a congressman, was indicted on federal corruption charges for taking campaign funds for personal use. He was convicted of ten felony counts including bribery, racketeering, and tax evasion.

Dan Rostenkowski, a congressman, was indicted on corruption charges in the House Post Office scandal. He was charged with keeping "ghost" employees on his payroll and using Congressional funds to buy gifts, diverting taxpayer funds to pay for vehicles for personal use, tampering with grand jury witnesses, and trading officially purchased stamps for cash at the House Post Office. He was fined and sentenced to seventeen months in prison.

Abscam, a Federal Bureau of Investigation sting operation in the 1970s and 1980s targeted trafficking in stolen property and corruption of businessmen resulted in the conviction of Congressmen Harrison Williams, John Jenrette, Richard Kelly, Raymond Lederer, Michael Myers, Frank Thompson, and John Murphy. They all received various prison sentences.

All of those above mentioned were duly elected to lead our country as legislators. Imagine that! This seems to support the need for government financed elections, so that corrupt individuals cannot buy their way into government leadership as a means of covering their criminal activities.

My old friend, Skip Sleyster, would have said: …"these guys got caught with their hands in the cookie jar."

CHAPTER **19**

Governmental Choices

THE CHOICES OF forms of government vary from monarchy to communism, but ours is more or less a quasi-democracy. We call it a democracy but considering our senatorial representation which disregards population, it renders each state of equal power in the Senate, an undemocratic principle. The voter's choice of principals of government is primarily between the Democratic or Republican parties.

The Democratic Party, generally considered the party of the people, is thought to be favored by the "working-class" of voters, those voters who focus on family values and economic equality. Democrats are considered to be on the left side of politics. In politics, the term left refers to people and groups that have liberal views. That generally means they support progressive reforms, especially those seeking greater social and economic equality. The party is thought by some to favor extreme views, such as communism and socialism, although such is not the case by any stretch of intelligent imagination.

The Republican Party is generally considered as the party of the upper class and corporate executive class of voters, those voters who focus on management and ownership of the tools of production and distribution of the goods and services. Those of these political ideals, with more conservative views, are considered to be on the right side of politics. They prefer to resist or limit change and favor governmental control in the hands of the elite who, in their opinion, keep the

GOVERNMENTAL CHOICES

economy working for everyone.

The origin of the political left and right do actually have reference to the physical directions, left and right. Left and right originally referred to seating positions in the 1789 French National Assembly, the parliament France formed after the French Revolution.

The forms of government are many. The various forms and basic different roles of leadership power include the following classifications: [84]

- **Democracy** is a form of government that allows the people to choose leadership. Democracies advocate for fair and free elections, civic participation, human rights protections, and law and order.
- **Communism** is a centralized form of government led by a single party that is often authoritarian in its rule.
- **Socialism** is a system that encourages cooperation rather than competition among citizens. Citizens communally own the means of production and distribution of goods and services.
- **Oligarchies** are governments in which a collection of individuals rules over a nation. A specific set of qualities, such as wealth, heredity, and race, are used to give a small group of people power
- **Aristocracy** refers to a government form in which a small, elite ruling class — the aristocrats — have power over those in lower socioeconomic strata. Members of the aristocracy are usually chosen based on their education, upbringing, and genetic or family history.
- **Monarchy** is a power system that appoints a person as head of state for life or until abdication. Authority traditionally passes down through a succession line related to one's bloodline and birth order within the ruling royal family, often limited by gender.
- **Theocracy** refers to a form of government in which a specific religious ideology determines the leadership, laws, and customs.

141

- **Colonialism** is a form of government in which a nation extends its sovereignty over other territories. In other words, it involves the expansion of a nation's rule beyond its borders.
- **Totalitarianism** is an authoritarian form of government in which the ruling party recognizes no limitations whatsoever on its power, including in its citizens' lives or rights. A single figure often holds power and maintains authority.
- **Military dictatorship** is a nation ruled by a single authority with absolute power and no democratic process. The head of state typically comes to power in a time of upheavals, such as high unemployment rates or civil unrest.

The most common forms of government include: [85]

- **Presidential republic:** Head of state is the head of government and is independent of legislature
- **Semi-presidential republic:** Head of state has some executive powers and is independent of legislature; remaining executive power is vested in ministry that is subject to parliamentary confidence
- **Parliamentary republic with an executive presidency nominated by or elected by the legislature:** President is both head of state and government; ministry, including the president, may or may not be subject to parliamentary confidence
- **Parliamentary republic with a ceremonial presidency:** Head of state is ceremonial; ministry is subject to parliamentary confidence
- **Constitutional monarchy:** Head of state is executive; Monarch personally exercises power in concert with other institutions
- **Constitutional parliamentary monarchy:** Head of state is ceremonial; ministry is subject to parliamentary confidence
- **Absolute monarchy:** Head of state is executive; all authority vested in absolute monarch

- **One-party state:** Head of state is executive or ceremonial; power constitutionally linked to a single political movement
- **No constitutionally defined basis to current regime**

The United States is considered a Federation, which is basically defined as a state in which the federal government shares power with regional governments with which it has legal or constitutional parity. [86]

The questions of the effectiveness, fairness, equality of opportunity, equality of law enforcement, redress of grievances, protection from financial misrepresentation, equality in penalty for criminality, among others, are paramount in our federation, which we consider a system of quasi-democracy. These questionable measures of equality can all provide evidence of the influence of the power of money on the disposition of claims or charges of wrongdoing or criminality. [86]

It would be literally impossible for a person unrepresented or represented by a public defender to receive the same level of representation as a person who has the financial ability to be represented by private counsel of choice. The caseload of a public defender would typically be onerous compared to that of privately-hired counsel. Such a situation would tend to lead public counsel to take the easy road to settlement in favor of the rich and powerful instead of engaging in lengthy preparation of defense for the unrepresented accused. Just the cost of expert testimony in such litigation would be a disadvantage for the underfunded party. Equality cannot prevail under such a system.

So, in consideration of all the descriptions in classifying forms of government, the United States is said to be democratic, however, one could question such a narrow classification due to the electoral process being dominated by campaign finance, most of which is provided by the wealthy individuals and corporations. Such a process is influenced by the power of money in the electoral process which is provided primarily by the wealthy individuals and corporations rather than the populous in general. This creates the inference of an undemocratic influence of the rich and powerful in electing representatives to serve their needs and wishes rather than the needs and wishes of all Americans.

CHAPTER **20**

Alternative Campaign Finance

LEGISLATING A BAN on campaign finance by individuals and corporations will be a challenge with enormous opposition by those who have their foot on the pedal of influencing governmental actions for personal advantage, as well as those who have idealistic beliefs regarding freedom of speech. This challenge became much more prominent with the "Citizens United" Supreme Court decision declaring that corporations may provide financial influence in the electoral process, a decision which more or less declares that corporations are people.

This seems to fly in the face of the original intent of legislation regulating the electoral process which declares that each citizen shall have one vote in any election. Allowing corporations to influence the electoral process gives corporation owners additional influence in an election through the power of money, an opportunity that most individuals do not have.

This process has been heavily debated for decades without any solution which satisfies the powerless citizen lacking funds to counter the corporate influence of money.

Perhaps the only viable solution will require a constitutional amendment dealing specifically with campaign finance, one without potential controversy of interpretation, if that is possible. Such an amendment could ban individual or corporate financial support of any candidate for office. Perhaps each candidate should also be banned from personally funding a campaign. And perhaps campaign advertising should be provided free of charge by any publication for any candidate a limited amount not considered a financial hardship on the given publication. And perhaps any publication should be banned from expressing an opinion about a given candidate for office. The power of the press is substantial. Their potential influence on voting creates an extremely undemocratic electoral process, one in which equality of opportunity is impossible to achieve.

The problem with information regarding candidates for office is that there is no process of verifying claims about a candidate's background or principles without the financial wherewithal to perform the steps necessary for verification or rejection of such claims or beliefs. There is a cliché which says that if you tell a lie often enough you tend to believe it yourself. The only time it becomes evident that someone's claim of political persuasion is not as declared is after they are elected to office and then make decisions which differ from their declared intent.

It would seem unreasonable to expect an elected official to ignore the influence of their campaign financiers in the decision-making process regarding legislative or enforcement matters that affect such financiers' beliefs and personal affectations. Another quotation states that you should dance with the one who brought you to the party, an important thing for a candidate to remember since a reelection campaign initiates the fundraising process all over again.

It would seem imperative in considering the candidate to support, to check the sources of campaign funding in the current or previous campaigns of such candidate. Another worthwhile source of a candidate's political persuasions can be partially determined by detailed information about campaign funding sources – again, political philosophies are meaningful, but beyond the basics of progressiveness or conservativeness, many elected officials tend to dance with the one who brought them to the party, and not necessarily because they really desire to, but because it is essential to assure remaining in office.

ALTERNATIVE CAMPAIGN FINANCE

Needless to say, we cannot have a democratic government which serves on a basis of equality of opportunity and privilege as long as election campaigns are financed privately. Human nature, political motivation, retention of standard of living, career enhancement, friendships, relationships, and other similar considerations stand in the way of devotion to duty and equality of opportunity considerations when it comes to retention of one's political future. That will not change unless campaign finance changes. So, unless that takes place, we will have a government bought and paid for by the rich and powerful individuals and corporations. They pay for the power and privilege and they are going to have it.

CHAPTER **21**

Taxation – the only path to equality

TAXATION SEEMS TO be the only viable solution to a administering a reasonable measure of equality of income and wealth. Not to eliminate the economic value of the quest for higher income and wealth in enabling capitalism to function as the driving force of equality of opportunity, but as a means of providing limitations on financial exploitation by those who have mastered positions of power and privilege.

Equitable taxation should be considered taxation based upon the ability to pay, a simple system structured to be fair to those of all levels of earnings. The system in the U.S. is considered progressive taxation, which imposes a lower tax rate on low-income earners than on those with a higher income. This is usually achieved by creating tax brackets that group taxpayers by income ranges.

The U.S. system is considered a progressive system, although it has been growing flatter in recent decades. For 2021, there are only seven tax brackets, with rates of 10%, 12%, 22%, 24%, 32%, 35% and 37%. There previously were 16 brackets.

The rationale for a progressive tax is that a flat percentage tax would be a disproportionate burden for people with low incomes. The dollar amount owed may be smaller, but the adverse effect on their real spending power would be greater.

The rational for a progressive tax system is that it reduces the tax burden on the people who can least afford to pay. That leaves more money in the pockets of low-wage earners for the purchase of essential goods and services which ultimately stimulates the economy in the process.

A progressive tax system also tends to collect more taxes than flat taxes or regressive taxes, as the highest percentage of taxes is collected from the highest amounts of money.

A progressive tax also requires those with the greatest amount of resources to fund a greater portion of the services that all citizens and businesses rely upon, such as road maintenance and public safety.

Critics of progressive taxes consider them to be a disincentive to success. They also oppose the system as a means of income redistribution, which they believe punishes the rich, and even the middle class, unfairly.

Opponents of the progressive tax generally are supporters of low taxes and correspondingly minimal government services.

The opposite of a progressive tax, a regressive tax, takes a larger chunk of disposable income from low-wage earners than from high-wage earners.

A sales tax is an example of a regressive tax. If two individuals, one rich and one poor, buy an identical bag of groceries, both pay the same amount of sales tax. But the poorer person has shelled out a greater percentage of his or her income in order to get those groceries.

A flat income tax system imposes the same percentage tax on everyone regardless of income. In the U.S., the payroll tax that funds Social Security and Medicare is often considered a flat tax because all wage earners pay the same percentage. However, this tax has a cap. For 2021, the payroll tax is not applied to earnings over $142,800. (For 2022, the tax is not applied to earnings over $147,000). This makes it a flat tax only on the people who earn less than that amount. Taxpayers earning more than $142,800 a year pay a lower percentage of their overall income in payroll taxes. That makes it a regressive tax.

So, is our system progressive, regressive, or a flat tax? Well, as detailed above, it has elements of all the above. Our income tax is progressive, the higher one's income the greater the tax. It is regressive in a sense because the payroll tax ends at some point with higher incomes. It is considered a flat tax because every level of income earned requires the same percentage for payroll taxes.

One's tax bracket depends upon taxable income and the filing status: single, married filing jointly or qualifying widow(er), married filing separately and head of household. Generally, as you move up the pay scale, you also move up the tax scale.

The IRS recently announced new tax brackets for the 2022 tax year, for taxes you'll file in April 2023, or October 2023 if you file an extension. There are seven tax brackets for most ordinary income for the 2022 tax year: 10%, 12%, 22%, 24%, 32%, 35% and 37%.

It should be noted that during the Great Depression in the early 20th Century the income tax rate reached 90% on high income earners. The Franklin D. Roosevelt administration considered it necessary for the sake of economic survival that the burden should be on the high income earners to save the economy which enabled them to prosper and that it is deemed imperative to save the wage-earners from complete financial disaster.

Who or what system establishes the pricing of goods and services? Who or what system establishes the value of the services or functions of various jobs or professions? Who are what system decides the economic value of the services of a trash collector or a common laborer without special skills? Who decides the value of the services of a school teacher, a healthcare worker, a policeman, a fireman, a railroad engineer, a carpenter, a mason, a painter, a childcare worker, a parking attendant, a garbage collector, a secretary, an accountant, an office assistant, a grocery clerk, a municipal laborer, an attorney, a medical doctor, or any of the myriad of other professions or jobs that are performed each day?

Is it decided by someone of psychic intelligence who can weigh the value of each and every service or function of society that is required or enjoyed by humanity? Is a doctor more valuable than a garbage collector or a school teacher? Well, it depends upon what one needs at the given moment. Is a lawyer more valuable than a carpenter or a mason? Again, it depends upon what one needs at a given moment.

It is obvious that the value of a given worker varies continuously by the need of the moment. Of course, some professions require much more education, training, and experience for the individual to become proficient. Such requirements could be considered as justifiable measures to increase the value of the provider over those of professions requiring lesser skills, knowledge, or experience. But, who decides all this?

It is decided to some degree by the system of supply and demand, the demand for such services and the number of available providers of such services. But it is also decided by collusion on the part of the body of providers of such through mutual agreement of those who are able to provide the services. This occurs through labor unions, associations of providers, or the laws of supply a demand. If your house is on fire you don't seek the lowest bidder to put out the fire. If you are sick or injured you don't interview available doctors or clinics to cure what ails you. If you want to buy a car you can certainly shop around but if you need groceries you don't interview the local markets to determine where the lowest wages are paid to the employees.

Basically, those who have special skills or educations for various professions tend to value their services by the same method which determines the value of a common laborer - supply and demand. This practice, which seems appropriate, is more common with those less-skilled than those with special skills. Common laborers don't have associations whereby they can establish minimum rates for their services but professional do. Lawyers have bar associations whereby collusion is established, doctors have medical associations whereby collusion is established, but school teachers lack equivalent methods of establishing equality of charges for their services. This is particularly noteworthy since there is no profession more imperative for the future of mankind than the teaching profession. Without that it seems that there would be no other profession.

It is obvious that the power to establish the value of one's services focuses of the title of this writing – power and privilege. The only way to limit the negative impact of such an unequal method of remunerations for various services is through taxation. The taxation of income should be the method of leveling this playing field. It is the only method available. It is a practice abhorred by most of those with higher income however some openly suggest that they be taxed more than they are.

Attempts to level this playing field of income and wealth of course creates the profession of the political mastermind. It keeps the Republican Party alive and thriving. And it keeps the Democratic Party alive and thriving. The Republican Party is commonly thought to be the party preferred by the rich and powerful, the corporate or wealthy class, and the Democratic Party is commonly thought to be the party preferred by the working-class of wage and salaried workers. It seems that many people tend to declare their allegiance to the party emphasized in their family relationships regardless of their financial wherewithal.

CHAPTER **22**

Funding of Political Parties

THE DESIRE TO select the candidate of one's choice must be enormous amongst the rich and powerful. The money raised and spent grows with each election cycle. The rewards must be beyond comprehension. The desire for fame, fortune, power, and privilege seem to attract excess fortunes into circulation, not for the good of humanity, but for personal advantage of some form. Otherwise, this unneeded wealth could go to helping those who were used without adequate remuneration in assembling such wealth for the ungrateful and clever acquirers.

It seems impossible to garner excess wealth while operating in a level playing field. It would seem to require inordinate remuneration for the goods and services of those clever enough to structure the business environment to favor their chosen profession or business endeavor.

Where does all the campaign funding money come from?

Funding available to the political parties during the 2020 elections: [59]

2020 Election Cycle	Total Raised	Total Spent
Democratic Party	$1,808,559,937	$1,737,081,592
Republican Party	$2,003,241,832	$1,931,908,122
Democratic Committee	$492,683,091	$462,362,409
Republican Committee	$890,538,963	$833,510,910
Totals	**$5,195,023,823**	**$4,964,863,033**

It would appear that funding and spending was about 25% greater by the Republican Party than the Democratic Party in the 2020 election cycle. The Republican Party is traditionally more likely to be funded by the upper income segment of society and the Democratic Party is traditionally more likely to be funded by the working-class segment of society.

Apparently, campaign coordinators are not satisfied with the news accounts of the qualities and political standings of their candidates. They are compelled to spend outlandish amounts of money raised from those who have a motive in funding their chosen candidates. Five billion dollars would have housed about five hundred thousand homeless people for one year. Meanwhile, the news media, without the lavish political spending by campaign promoters, would have fully informed the voters of the qualities and political persuasions of the candidates for president of the country.

The impact of five billion dollars to sway the voters is questionable at best. Political prognosticators tell us that most voters tend to follow party lines in selecting their choice of the candidate to support. The number of minds changed by the five billion dollars, although indeterminate, is unlikely to have swayed the election results. But at least it served to get some of the unneeded wealth of the rich and powerful back into circulation in the economy. Perhaps that is the major benefit of campaign funding.

Political Party Funding is a method used by a political party to raise money for campaigns and routine activities. The funding of political parties is an aspect of campaign finance.

Political parties are funded by contributions from multiple sources. One of the largest sources of funding comes from party members and individual supporters through membership fees, subscriptions and small donations. This type of funding is often referred to as grassroots funding or support. Solicitation of larger donations from wealthy individuals, often referred to as plutocratic funding, is also a common method of securing funds. Parties can also be funded by organizations that share their political views, such as unions, political action committees, or organizations that seek to benefit from the party's policies. In certain locales, taxpayer money may be given to a party by the federal government. This is accomplished through state aid grants, government, or public funding. Additionally, political fundraising can occur via illegal means, such as influence peddling, graft, extortion, kickbacks and embezzlement.

The ten largest individual political contributions in 2019 and 2020:

Contributions in 2019-2020 Campaign [87]	
Sheldon G. Adelson	$218,168,500
Michael R. Bloomberg	$152,509,750
Thomas Steyer	$72,119,974
Richard Uihlein	$68,314,982
Kenneth C. Griffin	$67,423,384
Timothy Mello	$60,097,555
Dustin Moskovitz	$50,568,012
Stephen A. Schwarzman	$35,466,600
Karla Jurvetson	$33,796,890
Jeffrey S. Yass	$30,581,500
James H. Simons	$26,320,791

One could certainly assume that the desire for power and privilege is a factor for those noted in the above examples of individual contributions in 2019 and 2020. Contributions given by those cited above to assist those in need or those who lack influence in the governing process are not listed as a comparison of their generosity.

The average income of an individual taxpayer in his or her lifetime would yield a mere fraction of an amount available for a political contribution of any of the above examples of this undue weight on the scales of democracy. The above examples are of contributions from one year for those listed. Our only way of leveling this inequality on the influence of government can only be leveled through taxation policy. One could assume that the intent of these egregious political contributions by the wealthy is to assure that the taxation policies are left unchanged so that the level of inequality persists for the rich and powerful?

Sheldon Adelman is said to have worth of $33 billion, a fortune mostly earned in the casino industry whereby gamblers are lured by the slim chance to win and are cleverly separated from their money.

Michael Bloomberg is said to have worth of $55 billion, a fortune mostly earned in the securities industry whereby those who own the brokerage businesses make a fortune by separating investors from their money.

Tom Steyer is said to have worth of more than $1 billion dollars, a fortune mostly earned in the hedge fund industry whereby hedge fund creators separate investors from their money with risky gambles on securities values.

It is obvious that anyone having financial worth in the billions didn't make their money working overtime on an assembly line or any other normal employment opportunity. The only explicable assumption is that those garnering higher income have some unfair advantage which enables extraordinary profits from their businesses or professions.

Any method of employment is structured to provide the employed with an ample standard of living, some with opportunities to earn higher than normal incomes. However, employment situations would rarely, if ever, provide an opportunity to amass a fortune such as the wealthy mentioned above possess and therefor would render extraordinary political contributions unrealistic.

Most fortunes come about from clever management of funding available initially from family wealth followed by good reinvestment and management, or from ground-breaking invention or development which enables protection from competition such as patent or copyright restrictions. Fortune also comes about from appreciation of the value of real estate due to location, development opportunities, or restrictions of like-kind development.

CHAPTER **23**

Income Tax Data

IT IS INTERESTING to note the distribution of income and wealth in our so-called "free market economy." When considering who owns and controls the marketplace it becomes obvious that the free market isn't free. That term is just a charade to placate the populace with a feeling of honesty and equality, a system which is the haven for those who know how to utilize the power of the media for personal gain.

"Free market is defined by Britannica.com as" an unregulated system of economic exchange, in which taxes, quality controls, quotas, tariffs, and other forms of centralized economic interventions by government either do not exist or are minimal. [88]

Many economists consider resource allocation in a free market to exist where no one can be made better off without making other individuals worse off. And, according to this theory, through the invisible-hand mechanism of self-regulating behavior, society benefits by having self-interested actors make free economic decisions that benefit themselves. Some ethicists have argued that the efficiency of free markets depends on conditions such as fair play, prudence, self-restraint, competition among equal parties, and cooperation.

Critics of the free market system tend to argue that certain market failures require government intervention. First, prices may not fully reflect the costs or benefits of certain goods or services, especially costs to the environment. Public goods are often underinvested or exploited to the detriment of others or future generations, unless such exploitation is prohibited through government regulation. Second, a free market may tempt competitors to collude, which makes antitrust legislation necessary. Antitrust and similar regulations are especially necessary in cases where certain market actors, such as companies, have acquired enormous market power. Third, transaction costs may mean that some exchanges are best performed in a hierarchy rather than in spot markets.

In response to these critiques, economists Ronald Coase, Milton Friedman, Ludwig von Mises, and Friedrich A. Hayek, among others, have argued for the robustness of markets because they can adjust to or internalize supposed market failures in many situations. For instance, many goods traditionally conceptualized as public goods requiring government provision have been shown to be open to free market contracting. Libertarians are strong defenders of the idea that a system of free markets provides the best economic system." [88]

Income tax data provides an indication of the distribution of income in the U.S.

Internal Revenue Service data on individual income taxes for tax year 2018:

In 2018, 144.3 million taxpayers reported earning $11.6 trillion in adjusted gross income and paid $1.5 trillion in individual income taxes.

Since 2001, the share of federal income taxes paid by the top 1 percent increased from 33.2 percent to a new high of 40.1 percent in 2018.

In 2018, the top 50 percent of all taxpayers paid 97.1 percent of all individual income taxes, while the bottom 50 percent paid the remaining 2.9 percent.

INCOME TAX DATA

The top 1 percent paid a greater share of individual income taxes (40.1 percent) than the bottom 90 percent combined (28.6 percent).

The top 1 percent of taxpayers paid a 25.4 percent average individual income tax rate, which is more than seven times higher than taxpayers in the bottom 50 percent (3.4 percent).

The above data indicates that the average tax rate on incomes was about 13% and the top 1% of earners payed about 40% of such taxes. If the top 1% of earners paid an average tax rate of 25% they must have earned more than $2 trillion, an average of more than $1 million each.

Taking such an equation a step further in justifying our tax fairness, the top 1% of taxpayers had an average income of about $750 thousand, half had more than $750 thousand and half had less than $750 thousand, after tax to support their lifestyles. The lower 99% of taxpayers had an average of about $60,000 after tax - half had more than $60 thousand and half had less than $60 thousand to support their individual or family lifestyles.

The above analysis of data is not only meant to be exemplary of the fairness or lack of fairness of our system of taxation, a system which is enacted and protected primarily by the political interests of those who support our elected officials in the U.S. government. It also is meant to suggest that those individuals and corporations with higher income and wealth can and should pay a larger share of the federal budget. This would render a larger share of the income of the wage-earner class, which includes most of us, available to provide additional market segment for the benefit of the total economy including those with extraordinary income and wealth. Perhaps this could be a win-win for all taxpayers.

But, like the saying goes, some things never change. Pandemics are not all bad. Just ask the billionaires of the world.

"Billionaires' wealth surges to record during pandemic" [89]

The share of global wealth held by billionaires surged to a record during the Covid-19 crisis, according to a group founded by French economist Thomas Piketty.

"About 2,750 billionaires control 3.5% of the world's wealth, the Paris-based Global Inequality Lab said in a report Tuesday. That's up from 1% in 1995, with the fastest gains coming since the pandemic hit, the group said. The poorest half of the planet's population owns about 2% of its riches."

"The study's findings add to a debate about worsening inequality during a public health crisis that's hurt developing economies – which are short of vaccines as well as financial resources to cushion the blow – even more than advanced ones. Within the rich world too, financial and real-estate markets have soared since the depths of the slump last year, widening domestic gaps."

"Those pandemic trends come after decades of policy that was often geared toward people at the top, on the expectation that it would "trickle down" and everyone else would ultimately benefit too, according to Lucas Chancel, co-director of the World Inequality Lab, said in an interview. "There is really this polarization on top of a world that was already very unequal before the pandemic," He said billionaires accumulated 3.6 trillion euros ($4.1 trillion) of wealth during a crisis in which the World Bank estimates that some 100 million people have fallen into extreme poverty."

"Across most parts of the world, the richest 10% of people control roughly 60% to 80% of wealth. Overall, poorer countries have been catching up with richer ones – but within those developing nations, inequality has soared. Same-country disparities now account for more than two-thirds of global inequality, up from roughly half in 2000, according to the Lab." [89]

This is another reminder that Reagan's "trickle-down theory" never worked and never will. Wealth doesn't trickle down, it rises up and multiplies as the rich get richer and the poor struggle along. When the

INCOME TAX DATA

owners of the wealth get wealthier they don't pass it on to the workers. They pile it up and savor it like vultures over a corpse. (Author comment)

So, what can we do to provide more equality of income and wealth in our democracy turned plutocracy? As stated above the federal government reveals that in 2018, the top 50 percent of all taxpayers paid 97.1 percent of all individual income taxes, while the bottom 50 percent paid the remaining 2.9 percent. The reason for that is obvious, the top 50 percent of taxpayers must have earned about 97 percent of the income in the country and bottom 50 percent must have earned about 3 percent of the income in the country. What tax policy would seem more just? It stands to reason that if the high income earners paid most of the tax they made most of the income. What the information available doesn't tell us is how much income and wealth did the rich and powerful have left after tax. We can only extrapolate that equation by considering the wealth of those with such high income.

The only fair and balanced method of assuring an economy with resilience for future generations would seem to tax wealth. Most wealth is established by appreciation of investment values or extraordinary income from such investments or industries. Extraordinary income is the definition of extraordinary profit margins for the goods and services provided. This extraordinary profit should be returned to the economy that provided it so that it will sustain the bountiful marketplace which the workers provide for the rich and powerful. In other words, these earnings enjoyed by the rich and powerful, which is provided by the working class, should be returned to the financial system to provide sustenance for the system to prevail for future generations to come.

In other words: tax the rich! The rich have a bill to pay. They owe the rest of us for providing an economy which enabled their riches. Now they should pay this debt amassed through favorable tax laws which enable avoidance of a fair tax impact on huge income and wealth to pay the debt they owe to society. It is long overdue.

CHAPTER **24**

The Ultimate Challenge

BEYOND FINANCIAL EQUALITY and the persistent necessity to provide a more fair and equitable system of taxation, lays another more important consideration. The rich and powerful, who are major contributors to the ultimate demise of planetary conditions which support human life, must lead the way in the necessity to change our actions which endanger the ability of the planet to continue to support human life. This need must take precedence over financial considerations. If we don't protect the life-sustaining ability of the planet then financial considerations such as equality of opportunity in income and wealth are useless goals to pursue. Ignoring our care of the planet in favor of financial success alone could lead to perilous conditions which would be not tolerable or reversible.

This all-important course of action takes precedence over any thoughts of financial equality. The recent book by this author: "Aces & Eights- Gambling on Climate Change" – published by Outskirts Press in 2020, stated that the ultimate importance of the precedence of changing our lifestyles must be the primary concern for mankind.

The book stated that: "The subject of climate change is a question of population control and planetary protective measures to reduce the waste and destruction of our natural resources by reducing our all-intensive pursuit of pleasure. This is not something about which we can say: "I'm thinking about it."

THE ULTIMATE CHALLENGE

The question posed is: Are we gambling with our living conditions and those of future generations with a losing strategy?

And we might add, are we allowing those with power and privilege to destroy our planet with their excess wealth garnered by political influence purchased with the power of money?

The song line we used to sing on the way to Boy Scout camp in the summertime as young teenagers, "You can't get to heaven on roller skates," may come into play here. Perhaps roller skates or bikes or walking would be a more astute choice of transportation if we intend to cease our continuous increasing of the belt of carbon dioxide with which we have encircled the plant and which traps carbon dioxide in our atmosphere and slowly heats the planet.

And if the United Nations could suddenly become a world body which focuses on saving the planet, perhaps financial considerations could take second place and some positive measures for planetary survival could become the number one issue of the day.

If we don't assure the world populace of a viable planet for human habitation, the most essential requirement for sustenance of humanity, then taxing the rich to stabilize a fragile economy won't matter for future generations of consumers. But it must be done here and elsewhere around the planet. If the too-late date has not passed it is not far off. The time for positive action is not in the future, it is now.

Perhaps "Save the economy: Tax the rich," or "Save the planet: Tax Carbon" are slogans of mutual necessities and of equal importance. They are both imperative for action for all inhabitants of Planet Earth. The longer we wait to reduce carbon from our atmosphere instead of increasing it, the shorter may be the endurance of the human race.

All of these facts and figures tend to support the pressing question of the day: "Can we continue to allow power, privilege, and the pursuit of pleasure to take precedence over sustaining the ability of Planet Earth to support human life?" This is the question upon which we must focus. The best time to do so was long ago but the next best time to do so is now.

The alarm bells are becoming more frequent by print media

correspondents. The warnings displayed in actual incidents make headlines almost daily in attempting to instill focus on the burden our lifestyles are impacting on the future of humanity.

Some recent examples:

Los Angeles Times, December 12, 2021, "Chaos in the High Seas in a Time of Climate Migration," by Ian Urbina.

"Climate change is expected to displace 150 million people across the globe in the next 50 years. Rising seas, desertification, and famine will drive the desperate to places like Europe and the U.S.

Urbina published a photo of a rubber raft type vessel with an outboard motor operated by Doctors Without Borders containing dozens of migrants from Bangladesh heading from Libya to Europe. This is just one example of desperate individuals willing to fork over every dollar they can muster and abandon any belongings they left behind to seek an opportunity for a livable environment to live their lives.

As it turned out, the captain of this voyage which was destined for disaster was taking these victims back to a gulag of migrant detention facilities in Libya where they would face possible rape, extortion, forced labor, and death.

This is but one example of the desperation of migrants to seek a better life and the thoughtless and mindless resolve offered by countries who are capable of accepting such migrants and offering them opportunities for a better life.

A recent article in Mother Jones magazine stresses the point: "Suck it up" - **"A Machine that farms the sky"** Clive Thompson, Mother Jones, December 2021. [91]

Clive Thompson tells us of an industrial plant in a valley of the Squamish River in British Columbia which is designed to suck carbon dioxide out of the air. It was built by Carbon Engineering, a pioneer in a technology known as air capture. This machine sucks in air and then causes it to react with a liquid chemical which grabs hold of the carbon dioxide molecules. If this technology could be replicated in masse it could begin a recycling of carbon and perhaps reverse climate change by capturing the surplus atmospheric carbon and burying it deep in the Earth, which Thompson calls it rewinding the Industrial Revolution.

Steve Oldman suggests that this captured carbon dioxide could be used to convert into liquid fuels for cars, trucks, planes, and power plants. He states that: "We can't wait, we have to get on with decarbonizing now."

The cost would be enormous, perhaps $5 trillion per year for the rest of the century. The Intergovernmental Panel on Climate Change (IPPC) warned us that our climate situation could decline so rapidly that we may be left with little choice. Oldman states that: "When water was a problem with cholera and typhoid, governments worldwide built a water treatment infrastructure. Today we have an air problem, so we need an air-treatment infrastructure."

The IPCC reported in 2018 that if we wanted to keep the planet from warming by more 1.4 degrees Celsius, which is the goal of the Paris agreement on mitigating climate change, we need to slash atmospheric carbon dioxide dramatically. Planting forests would help. Shifting to renewables would help. But given humanity's embrace of wind and solar that we would have to start pulling carbon directly out of the atmosphere by 2100, the demand would be ten billion metric tons per year, equal to nearly a third of our current carbon dioxide output.

In 2021 more than 100 of the world's largest companies had pledged to get to net zero emissions by 2040.

The challenge to pull carbon from the air has resulted in extensive governmental efforts to provide necessary funding. Such efforts for compliance include federal government and state government tax credits for firms that can pull carbon out of the atmosphere and other tax credits and direct investments in carbon capture actions.

The consensus of many environmentalists leans toward fossil fuel emission control as the primary method of dealing with the devastating debacle of climate change. Direct air capture, a process for sucking carbon dioxide from the air, is highly touted and pursued by the scientific community but considered not to be a first tool of choice. Due to the massive cost and development of such a system we should first focus on renewable sources of energy, treating beaches with chemicals to absorb carbon dioxide, and growing plankton to metabolize it. It was stated that farms produce about 100 million tons per year of bio-waste that could be used for sequestration.

Concluding Comments

THIS WRITING HAS touched on the past, the present, and the future. We have the mistakes and progress of the past as lessons for the future. The future is obviously going to be different than the past as a result of climate change and the challenges that it lays before us. We have the good fortune of reliance on the experts in science and engineering to lead the way. Their guidance is imperative to meet the challenges of climate change and the endangerment and impoverishment which it could entail.

None of this is newly discovered or revealing information. It has been available for all to absorb and digest. It is disregarded or ignored by many of us as we meet the challenges of life and pursue our individual employments and enjoyments. We seem to acknowledge that the warnings of our scientists may be true but we see no way that we can contribute as individuals to the cause of positive action, so we keep doing what we are doing.

A recent Los Angeles Times article: **"Americans Fear Climate Change,"** points out that we should be alarmed at how out of step our government remains and how little state and federal leaders have done in the face of this escalating peril. It states that: "Instead of acting decisively to slash emissions, switch to renewable energy, and phase out fossil fuels production, our government is still stuck in the mud, even as greenhouse gas emissions roar back after a pandemic-induced lull."

We cannot continue to focus entirely on comfort and pleasure as we meet these challenges. The environmental issues which come about as a result of our carbon-dependency which we have sought and developed for several centuries are an endangerment to all life forms. This presents a responsibility for all of us. We can all make a difference if we are willing. It cannot be a selective process, it must be all-inclusive.

Will self-sacrifice come about? Will it be sufficient? Must the indulgence and positive actions necessary require compliance? Must they be enforced? These are the compelling questions. The answers seem apparent. This challenge must be accepted and pursued worldwide. The leaders of every nation must understand the dilemma and agree to take the necessary steps to alter actions and lifestyles necessary to reduce our negative impact on the ability of the planet to support human life. Can it be done? Will it be done? It would be easy to say that it can or will. But the answer depends upon cooperation worldwide with immediate action. It may be difficult to attain universal compliance in a timely manner with a course of action deemed necessary by our environmental scientists. The best time to start this process of actions would have been a century ago, as we have learned, but the next best time to adopt and institute measures to placate the obvious end result of centuries of uncontrolled and extraordinary wasteful consumption of goods and services and personal pleasures is now.

Has the too-late date already passed? We don't know, but we can take the first steps toward planetary survival if the world leaders can instill the desire and incentives for compliance and the willingness for enforcement of mandatory measures. Enforcement will have to be mandatory. Willingness and voluntarism will be hard to achieve in any country, even our country, let alone all countries of the world as we all strive for a better life..

Achieving any progress in meeting the challenges which face the future generations of earthlings depends upon leadership from democratically governed nations. Although we do not function as a true democracy in America in the true sense of the term, the importance of striving for such is as important as maintaining our quasi-democracy which serves all the people instead of one which functions at the whim of one or few leaders under autocratic or plutocratic rule.

We certainly are not as democratic as some of us would wish, but we do strive for majority rule. Attempts for minority rule are ever present, but to date we have been successful in avoiding devastating results from such challenges. By virtue of our divisions of power between the executive, legislative, and judicial branches of government, we have maintained a government which attempts to serve all citizens. If we can make progress in eliminating the power of money from undue influence on the electoral process we will be more assured of serving our democratic intent, otherwise the challenge will continue to be a danger which lurks for our American form of quasi-democracy.

Governments such as ours are more inclined and more capable of initiating cooperation among nations in facing environmental challenges which we will incur from this time forward. Nothing can be deemed more imminent or destructive for mankind than the self-inflicted climate change which we are now experiencing and which will exacerbate in the future unless we take actions to lessen the impact.

Environmental conditions are not altered immediately by negative or positive actions. The process of change takes place over time - not days or months or years, but centuries. The adversity which we have imposed upon the planetary environment has been going on since civilization began massive progression a few centuries ago and has expanded as the population has increased and our pursuit of comfortable and entertaining lifestyles has progressed. It seems that we have reached the tipping point whereby any delay in altering environmentally-damaging actions endangers survival of humanity as we now know and enjoy it.

Do we have the right to destroy the planet's ability to support humanity in the next century? This is, of course, a rhetorical question. But the answer seems apparent. We don't have that right.

As we consider the challenges to life we should examine the

process of determining the causes and solutions to perils which will lessen the impact. It seems that any urgency for action is of more concern for the elderly among us than the young. This is probably a natural order of things since the young are more concerned with education, financial issues, occupations, careers, and family needs such as housing, and healthcare.

It has been said that wisdom is wasted on the elderly and energy is wasted on the young. I disagree with both statements. Everyone is naturally focused on his or her individual concerns at their stage of life at the given time. There will always be ways to improve the functions of government. The younger generations must be mindful of this along with the ever-changing environmental issues which threaten mankind. Those of us who have attempted to do the right thing now hand it off the next generation of well-educated and mindful citizens.

What seems to lie in the way of this ultimate debacle becoming universally accepted and challenged is politics? Of course we expect our elected leaders, our politicians, to understand and anticipate these threats to humanity, but what gets in the way? It is apparent, politics gets in the way, and the quest for power and privilege gets in the way. Here, and in many countries of the free and unfree world, politics, power, and privilege get in the way. It seems that we are not compelled to take positive action until the pain is felt. The question to which we have no answer is whether we have the option to wait until the pain is felt to treat the imminent injury.

The citizenry must choose the most capable and considerate among us to lead the recognition of the challenges and to collectively administer the appropriate courses of action to placate adversity and encourage future progress in protecting the planet from infinite disaster which could render it unable to support life.

This reminds us of the cliché: "Things that can happen will happen." A statement that should be kept in mind as we charge forward in consuming and depleting life-sustaining necessities which will be mandatory for the continuity of humanity and all other life forms on Planet Earth.

CONCLUDING COMMENTS

This acknowledgement may seem like bad news, but it could be good news if all of us accept the facts as gospel and begin immediately to change our lifestyles in a positive way to begin the movement toward survival for future generations of humanity, as unlikely as it may seem.

We have numerous current examples of the burden we have placed on our planet to support our ever expanding and improving lifestyles. The damaging weather episodes are becoming more frequent and more severe each year. Hurricanes, tornadoes, floods, wildfires, intolerable temperatures, rising sea levels, diseases, fresh water issues, and restoration of food sources are all becoming greater challenges.

Considering the enormity of the perilous conditions which we have instilled upon human habitability of our planet, we cannot afford, for our own benefit as well as future generations, to ignore our plight without taking every conceivable action possible to repair the damage done and eliminate these causes which challenge our way of life going forward.

In looking at the threat of peril resulting from our lifestyles it becomes apparent that power, privilege, and the pursuit of pleasure play a significant part in the potential plight of humanity and that we need to take steps to placate the imminent danger which lurks.

It seems apparent that the funding required for positive action is available from the enormous excess wealth amassed by the rich and powerful individuals and corporations. The wealthy have a greater footprint than the workers. They have had a larger impact on the demise of livability hereon than the workers. They have the wealth as evidence. They have prospered to the greatest extent and therefore should accept the financial burden to the greatest extent possible through taxation of their wealth. Wealth cannot be created out of thin air. It doesn't just show up in your bank account. It comes from the market of goods and services and good fortune which is produced by the populous, so in times of need that is where it must be returned.

In order to assure a healthy economy, one which serves all levels of income and wealth, we should take heed of this lesson of U.S. history from that spoken by Theodore Roosevelt in 1905, more than a hundred years ago, warning against the transmission and/or inheritance of massive fortunes. The results of political control of economic policy decision-making through the power of money in campaign finance over more than a hundred years since has enabled an unhealthy distribution of income and wealth. This must be controlled through taxation, not for the intent of leveling the playing field by placing the burden on the rich and powerful, but to inject the excess wealth back into the general welfare to provide assurance of a stable economy which serves all levels of income and wealth.

Without controlling this inordinate accumulation of wealth, which is enabled by the favorable taxation of income for the rich and powerful, there is no path to assurance of adequate education and employment opportunities for the general population. Without a financially stable general population we cannot maintain a healthy economy simply because there would be no market for the goods and services available. This is not rocket science, it is simple arithmetic.

It's like the cliché about a horse and carriage and love and marriage mentioned in the outset of this writing: "You can't have one without the other."

Addendum

MANY ARTICLES PUBLISHED in the Los Angeles Times express the dangers and demands of climate change that should be noted for detailed information regarding this ultimate challenge for humanity. They include:

"Our unhealthy addiction to oil ruins another coastline" - October 5, 2021 Editorial. The Orange County, California coastline became the latest casualty of the nation's unhealthy dependence on oil as a pipeline connected to an oil platform released 126,000 gallons of crude oil ashore with clumps of crude containing dead fish and birds. The oil slick was said to be larger than the City of Santa Monica. The beaches affected were subject to closure for weeks or months.

There are 23 oil and gas drilling platforms located in federal waters off the California coastline. Environmentalists have long warned that this aging infrastructure poses a serious risk.

This is, of course, nothing new. In 2015, a pipeline along U.S. 101 broke and sent more than 100,000 gallons of oil into the coastline, killing more than 200 birds and more than 100 mammals.

"Vanishing salmon push tribes to brink" – October 5, 2021. In a normal year, Alaska natives prepare salmon to tide them through the winter months. This year, there are no fish, both king and chum salmon have dwindled to almost nothing. The state banned salmon fishing on the Yukon.

FOOTPRINTS

Those studying the catastrophe generally agree that human-caused climate change caused the river and sea warmed which altered the food chain. Many believe that commercial fishing has exacerbated the problem along with global warming.

"Delay is denial on the climate crisis" – October 31, 2021 - by Michael E. Mann, Distinguished professor of atmospheric science and director of the Earth System Science Center at Penn State Univdersity. He is author of : "The New Climate War. The Fight to Take Back our Planet."

"Adaptation, resilience and carbon capture suggest action but fail to address the scale of the problem."

The author states that: "One can no longer credibly deny that climate change is real, human-caused, and a threat to our civilization.

Adaptation and resilience are important. We must cope with the detrimental effects of climate change that are already baked in – coastal inundation and worse droughts, floods and other dangerous weather events. But if we fail to substantially reduce carbon emissions and stem the warming of the planet, we will exceed our collective adaptive capacity as a civilization."

"Stop: Climate disaster ahead," November14, 2021

"There's no time to waste. The U.S. must quit dithering and adopt tougher auto emission standards."

The article discusses the gulf between countries' pledges to cur planet-warming pollution and what they are actually doing. It states that even if the world leaders deliver on their promises that a lack of concrete actin between now and 2030 means emissions will still be nearly twice as high as what's needed to limit warming to 1.5 degrees Celsius and avoid catastrophic climate change.

"Warming planet's effects hit home" – December 12, 2021. "California is no stranger to disasters, and climate change is making it worse."

ADDENDUM

California wildfires caused more than 50% of Los Angeles County residents from going outside at some point between summer 2020 and summer 2021 because of concerns about breathing wildfire smoke. Heat storms are taking a toll on quality of life. More than 75% of respondents to a survey said that climate change is a threat to their well-being. Heat, drought, wildfires, and air pollution are made worse by the burning of fossil fuels.

The article states that "Many politicians talk about climate change ad a problem to be solved for future generations, as oppose to a disaster that is making the planer less livable for current generations."

"Chaos on the high seas in a time of climate migration" - December 12, 2021 - by Ian Urbina.

This article features more than a hundred Bangladeshi migrants stuffed together in a rubber raft with an outboard motor operated by Doctors Without Borders being rescued while attempting to migrate from Libya to Europe to seek a better life. Efforts of survival becoming intolerable for these people, they risk the danger and their life savings to attempt to migrate to countries where better opportunities exist.

The author of this article summarizes the message: "Countries surely have a right and duty to manage their borders, but the way the U.S. and the EU are handling waves of these migrants is ineffective and inhumane. Putting merchant ship captains in the middle of this crisis is hardly the solution. Worse still is outsourcing the problem to failed states like Libya where human rights abuses are a foregone conclusion."

"Hopes wane at tornado-ravaged Amazon facility." December 13, 2021.

This article states that at least 6 people in an Amazon facility were killed in a tornado that struck near St. Louis in the Midwest. The sides of a warehouse collapsed inward and the roof caved in. The walls were constructed out of 11 inch-thick concrete.

But the real problem is the president himself, who can't shake the cobwebs of the Judiciary Committee that held its biggest hearings in the same ornate caucus room where he met with Democrats on Thursday.

He is too in the weeds on process. He's so lost in the snows of yesteryear that he is continuing his Amtrak Joe nearly-every-weekend commute to Delaware, albeit with better wheels, trading in the train for Marine One.

We want the president to rise above it and be an inspirational figure. We don't want the incremental updates of his negotiations with Joe Manchin.

We want to see Covid under control. We want to see the sacred right to vote protected. We want the grocery shelves stocked with affordable milk and meat. We want a president who tells us that we will get through this and we will be stronger for it.

Joe Biden better Build Better or he won't be Back. If he doesn't turn it around, he has cleared the way to a Republican rout in this fall's midterms. And in 2024, who knows how bad it can get?

"Warm weather a crucial ingredient in tornado outbreak" - December 13, 2021.

It was December but the warm, moist air made it seem like springtime as an eastbound storm spawned tornadoes that killed dozens over five U.S. states. The ferocity and path length of tornadoes in the Midwest and South put them in a category of their own, meteorologists say. Scientists expect atypical weather in the winter to become more common as the planet warms. It stated that tornadoes typically lose energy in a matter of minutes, but in this case it took hours. This storm was moving at a speed of well over 50 miles per hour. Scientists cannot attribute these storms specifically to climate change but they have observed changes taking place to the basic ingredients of a thunderstorm as the planet warms.

"Rainfall, beavers in warming Artic," December 15, 2021.

"High of 100.4 degrees reported amid signs of rapid deterioration in northern polar region."

ADDENDUM

The article states that the Artic continues to deteriorate from global warming, changing so rapidly that scientists call it alarming in their annual Artic report card. "The trends are alarming and undeniable," said U.S. National Oceanic and Atmospheric Administration chief Rick Spinrad, representing the findings of 111 scientists from 12 countries at the American Geophysical Union conference.

This report card comes as the Artic is warming two to three times faster than the rest of the planet. The melting ice opens the door to more pressures countries wanting to exploit the area. Those living there must adapt to a ground that is getting softer as permafrost melts. The reach of the sea ice was the 12th lowest on record. The rarer thick sea ice had its second-lowest reach since researchers began keeping records in 1985.

"Americans fear climate change" – January 13, 2022

"But you wouldn't know it from our government's dawdling and weak-kneed response."

Is it any wonder that more Americans are now alarmed by global warming, with the horrific onslaught of fires, floods, heat waves, and other disasters that we have experienced in the last year alone?

The article states that it is alarming enough but is the out of step way that our government remains and how little the state and federal leaders have done in the face of an escalating emergency. Instead of acting decisively to slash emissions, switch to renewable energy, and phase our fossil fuel production, our government is still stuck in the mud, even U.S. greenhouse gas emissions roar back after a pandemic-induced lull.

It further states that "we are running out of time to avert catastrophic warming and we can't allow dawdling and weak-kneed politicians to jeopardize our future with inaction and half-measures that are increasingly at odds with public opinion and reality. Leader\s must stop treating the destruction of the planet as one pet issue among many. It is the defining threat of our time, and they should be jolted into action."

Those of us who favor clichés of the past as simple explanations of things to come would probably categorize this lack of "practicing what we preach" as "closing the barn door after the horse has gone." It seems that the barn door is opening and the horse is looking and waiting. It seems that our only survival tactic is to practice what we preach and eliminate carbon emissions from the equation as fast as humanly possible.

And so far as equality of opportunity and economic progress are concerned, no one has said it better than Paul Krugman in his article: **"Tax the Rich, Help America's Children."** [92] He states that: "Republicans will, of course, denounce whatever Democrats come out with. But there are three things you should know about both taxing the rich and helping children: They're very good ideas from an economic point of view. They're extremely popular. And they're very much in American tradition."

He further states: "Although the modern Republican Party is utterly committed to the proposition that low taxes on corporations and the rich are the key to economic success, there is no evidence that this is true, If anything, the historical correlation runs the other way. The U.S. economy grew faster during periods when taxes were relatively high than it did when they were low."

"America has had progressive income taxes and estate taxes since 1916. In 1905 Theodore Roosevelt argued that it was essential to prevent the inheritance and transmission in their entirety of fortunes swollen beyond all healthy limits.

"A modern U.S. politician who said anything similar would be accused of engaging in un-American class warfare." [92]

Notes and References

1. Homer, Greek Poet – https://biography.com/writer/homer — 2
2. The Iliad — 3
3. The Odyssey — 4
4. Ulysses – James Joyce, Irish Novelist — 5
5. Leo Tolstoy – Russian Playwright and Novelist — 6
6. War and Peace, Leo Tolstoy — 6
7. War and Peace, Epilogue, Part II — 8
8. War and Peace, Leo Tolstoy — 9
9. William Shakespeare, As You Like It, Act II — 10
10. Victor Hugo – French writer – Les Miserables — 11
11. https://ballotpedia.org -Ruth Bader Ginsburg, — 14
12. https://en.wikipadia/Woody_Guthrie" — 16
13. Leonard Cohen, Canadian singer/songwriter — 19
14. "Every Knows" - Leonard Cohen — 20
15. "Democracy" - Leonard Cohen — 21
16. "Anthem" - Leonard Cohen — 21
17. "Get Up Stand Up" Bob Marley — 22

18. "Redemption Song" Bob Marley — 23
19. https://en.wikipedia.org/wiki/John_Lennon — 24
20. "Imagine" John Lennon — 25
21. "Alice's Restaurant," "City of New Orleans" Arlo Guthrie — 26
22. "Alfie" Burt Bacharach — 27
23. Harriet Beecher Stowe – "Uncle Tom's Cabin" — 28
24: Anna Elizabeth Dickinson – "Women's Work and Wages" — 29
25. Robert Reich – "The System, Who Rigger It, How to Fix It" — 29
26. Noam Chomsky – "How the World Works" — 30
27. Thomas L. Friedman – "Hot, Flat and Crowded" — 32
28. Paul Krugman – "The Conscience of a Liberal" — 33
29. Thomas Sowell – "The Wealth and Poverty of Nations" — 34
30. Peter Drucker – "The Age of Discontinuity" — 35
31. Albert Gore – "Earth in the Balance" "Politics of Wealth" — 35
32. William Greider – "The Soul of Capitalism" — 37
33. Upton Sinclair – "The Jungle" — 38
34. Joseph Stiglitz – "Freefall" — 39
35. Charles Ferguson – "Predator Nation" — 39
36. (https://en.wikipedia.org/wiki/Oliver_Stone) — 41
37. Al Franken – "Tax the Rich!" — 42
38. https://en.wikipedia.org › wiki › Schindler's List — 43
39. Bob Dylan – Singer/Songwriter — 44
40. Paul Simon – Simon & Garfunkel – Singer/Songwriters — 46
41. Neil Young – Singer, Songwriter — 47

NOTES AND REFERENCES

42. Arlo Guthrie – Singer/Songwriter	48
43. Billy Joel – Singer/Songwriter	49
44. Willie Nelson – Singer/Songwriter	50
45. The Beatles – Singer/Songwriters	51
46. Johnny Cash - Singer/Songwriter	51
47. Stevie Wonder- Singer/Songwriter	52
48. Sam Cooke - Singer/Songwriter	53
49. https://www.bls.gov - U.S Bureau of Labor Statistics))	56
50. IRS Fact Sheet - The Examination (Audit) Process	59
51. Jeffery Sachs – "The Price of Civilization"	65
52. https://fitsmallbusiness.com/newspaper-advertising-costs	71
53. https://adage.com/article/campaign-trail/political-ad-spendinyear-reached-whopping-85-billion/2295646	72
54. https://www.pbsd.org/newshour/show/unprecedented-spending-for-2020-political-ads	72
55. https://www.speakeasypolitical.com/pricing/?gclid=EAIaIQobChMItqKf0OGa8wIVLD6tBh2-dQ2GEAAYASABEgICsvD_BwE	72
56. https://callhub.io/phone-canvassing/	73
57. https://www.opensecrets.org/news/2021/02/2020-cycle-cost-14p4-billion-doubling-16/	74
58. https://robocent.com/?gclid=EAIaIQobChMIiOaFg-qa8wIVYxh9Ch0SVQ6WEAAYASAAEgIBuvD_BwE	74
59. https://en.wikipedia.org/wiki/Canvassing	75
60. Robert Scheer – The Great American Stickup"	78
61. C. Wright Mills –"The Power Elite-The Higher Immorality"	84
62. Thom Hartman-"Screwed-Undeclared Against Middle Class"	85

63. https://en.wikipedia.org/wiki/Occupy_Wall_Street — 90

64. https://minimum-wage.procon.org/ — 92

65. Bill Moyers – "Moyers of Democracy" — 95

66. Jacob Hacker/Paul Pierson – "Winner Take All Politics" — 95

67. Nicholas Goldberg - Los Angeles Times: "Why on Earth shouldn't we tax the super-rich more?" — 98

68. Jason Lanier – "Who Owns the Future" — 101

69. https://www.statista.com/chart/19635/wealth-distribution-percentiles-in-the-us/ — 104

70. Katharina Buchholz - "The Top 10 Percent Own 70 Percent of U.S. Wealth" — 107

71. https://en.wikipedia.org/wiki/Distribution_of_wealth — 108

72. Credit Suisse Research Institute's "Global Wealth Databook", Table 3-1, published 2021 — 109

73. https://en.wikipedia.org/wiki/Distribution_of_wealth — 109

74. https://www.google.com/search? client=firefox-b-1-d&channel=cus5&q=How+has+the+Earth%27s+atmosphere+changed+over+time&sa=X&ved=2ahUKEwj9wdy4rdTzAhVtJDQIHZBGBmQQ1QJ6BAgFEAE&biw=1200&bih=607&dpr=1.2 — 111

75. http://earthscience.stackexchange.com/questions/7644/is-oxygen-the-most-abundant-element-on-earth — 111

76. https://www.scientificamerican.com/article/how-the-environment-has-changed-since-the-first-earth-day-50-years-ago/ — 112

77. https://worldpopulationreview.com/country-rankings/democracy-countries — 122

NOTES AND REFERENCES

78. https://www.google.com/search?channel=cus5&client=firefox-b-1-d&q=what+is+the+entire+cost+of+our+legal+system   125

79. Read more: https://www.americanactionforum.org/research/the-economic-costs-of-the-u-s-criminal-justice-system/#ixzz7BHCyOAHu   125

80. Paul Krugman – "Tax the Rich, Help America's Children." October 25, 2021   131

81. Don't Know Much About History," Kenneth C. Davis   141

82. https://www.dictionary.com/e/leftright/?itm_source=parsely-api   142

83. https://thebestschools.org/magazine/common-forms-of-government-study-starters/   143

84. https://en.wikipedia.org/wiki/List_of_countries_by_system_of_government   155

85. https://www.opensecrets.org/2020-presidential-race   157

86. https://www.britannica.com/topic/auction   160

87. Augusta Saraiva and Alessandra Migliaccio – Bloomberg – December 8, 2021:   164

88. Los Angeles Times, December 12, 2021, "Chaos in the High Seas in a Time of Climate Migration," by Ian Urbina.   164

89. "A Machine that farms the sky" Clive Thompson, Mother Jones, December 2021   178

90. Paul Krugman - "Tax the Rich, Help America's Children."   178

www.ingramcontent.com/pod-product-compliance
Lightning Source LLC
Chambersburg PA
CBHW050211230526
45470CB00001B/334